SMM7 REVIEWED
DEARLE & HENDERSON

BSP PROFESSIONAL BOOKS

OXFORD LONDON EDINBURGH

BOSTON MELBOURNE

Copyright © Dearle & Henderson
1988

All rights reserved. No part of this publication may be reproduced, stored in a retrieval system, or transmitted, in any form or by any means, electronic, mechanical, photocopying, recording or otherwise without the prior permission of the copyright owner.

First published 1988
Reprinted 1989

British Library
Cataloguing in Publication Data
SMM7 reviewed.
 1. Great Britain. Buildings. Construction. Standard method of measurement
 I. Dearle & Henderson *(Firm)*
 692'.3

ISBN 0-632-02218-3

BSP Professional Books
A division of Blackwell Scientific
 Publications Ltd
Editorial offices:
Osney Mead, Oxford OX2 0EL
 (*Orders:* Tel. 0865 240201)
8 John Street, London WC1N 2ES
23 Ainslie Place, Edinburgh EH3 6AJ
3 Cambridge Center, Suite 208,
 Cambridge MA 02142, USA
107 Barry Street, Carlton,
 Victoria 3053, Australia

Printed and bound in Great Britain by
Hollen Street Press Ltd, Slough

Dearle & Henderson, chartered quantity surveyors, have grown steadily since their inception in 1908 and now comprise 18 partners and associates together with some 80 staff, working from six United Kingdom offices. Work in many parts of the world, including East and West Africa, the Middle East, Spain and the West Indies, is undertaken in conjunction with their associated practices overseas.

Complementing this continuing growth the range of skills has been expanded to include building services engineers, building surveyors and project managers. In addition to the more usual activities of quantity surveyors, Dearle & Henderson undertake research contracting, the preparation of large scale feasibility and planning studies, cost benefit analysis and life cycle costing, cost advice to manufacturers, the preparation of technical guides and text books, advice and assistance with the relocation of office and other business premises, energy surveys, maintenance programmes, project co-ordination and construction management.

Extensive use is made of microcomputer and electronic measuring equipment, both on site and in the office, for estimating, the preparation of tender and contract documents, financial control during construction and the preparation of final accounts, in addition to the more usual office functions.

Contents

Preface

Chapter 1 - Introduction and Brief Comparison with SMM6   1

Chapter 2 - Specification/Preambles   8

Chapter 3 - SMM6 to SMM7 : a Detailed Comparison   11

Chapter 4 - CESMM2 : How it compares with SMM7   276

Chapter 5 - POMI : How it compares with SMM7   324

Appendix A - List of SMM7 Work Group/Sections relative to Chapter 3   A/1-6

Appendix B - List of CESMM2 Work Classes relative to Chapter 4   B/1

Appendix C - List of POMI Sub-Sections relative to Chapter 5   C/1-3

Preface

The coming of the seventh edition of the Standard Method of Measurement (SMM7) with its new format was awaited with interest at Dearle and Henderson.

SMM7 has been prepared and should be used in conjunction with the other documentation published by the Building Project Information Committee. The other documents in the series are a Code of Procedure for Measurement of Building Works, a Common arrangement of work sections, a code of procedure for Production Drawings and a code of procedure for Project Specification.

After careful consideration a Working Party was set up within our practice to analyse the proposals in order that the requirements of SMM7 could be implemented promptly.

"SMM7 Reviewed" sets out the findings and suggestions of our working party and has been designed as an aid during the transition from the sixth to the seventh edition of the Standard Method of Measurement and beyond.

It was not our intention to draw conclusions regarding the possible advantages or disadvantages of the new documentation or to praise or criticise the content of SMM7. Our purpose in preparing this review was to provide an objective comparison with SMM6 and as subsidiary exercises comparisons with edition 2 of the Civil Engineering Standard Method of Measurement and with Principles of Measurement International.

We have made every effort to ensure that the information contained in this handbook is accurate but cannot accept any liability for loss incurred by the use of the information given.

ACKNOWLEDGEMENTS

The Partners of Dearle and Henderson wish to thank

C Strotton, N Kemp, R Holman, P Rickards, M Springett,

A Winter, J Armstrong and all others associated with

the preparation of this book

February 1988

Dearle & Henderson
Chartered Quantity Surveyors
4 Lygon Place
London SW1W 0JR

1 : Introduction and Brief Comparison with SMM6

The seventh edition of the Standard Method of Measurement of Building Works (SMM7) has been prepared in a tabulated format similar to that used in the Civil Engineering Standard Method of Measurement (CESMM2). When the document is first opened the tabulated format appears unfamiliar when compared with the prose of the sixth edition. However after a relatively short period the presentation was found by all members of our working party to be straightforward and logical. The rules are set up with a five column classification table with adjacent columns for measurement rules, definition rules, coverage rules and supplementary information. General information to be provided is always set out at the start of a section and horizontal lines are used between various classifications to inform the user which items should be zoned together.

Unlike previous Standard Methods of Measurement of Building Works SMM7 is not structured in trades but covers construction operations in the order set out by the Common arrangement of work sections for building works published by the Building Project Information Committee (BPIC). As previously stated the classification tables are set out in five columns. The first column gives a brief description of building works frequently encountered. The second and third columns give further breakdowns of the building activity and additional items can be added if the lists detailed in SMM7 do not show the precise requirement of the item to be measured. The fourth column sets out the unit of measurement. The fifth column details additional information that should also be stated if it applies. It is noted that more than one of the features listed in this column could apply and all of the features that do apply should be stated.

Measurement rules detail how the work item shall be measured and the method by which the quantities shall be calculated.

Definition rules to find the extent and limits of the work are represented by a word or expression used in the rules. If those words or expressions are used in Bills of Quantities then the definition rules also define them in that context.

Coverage rules draw attention to items of work which are deemed to be included. It is important to note that where coverage rules include materials they shall be mentioned in item descriptions.

The supplementary information column details further information which shall be given when a particular item is being described.

It is recommended that SMM7 is used in conjunction with the other documents published by the BPIC which are set out below with brief descriptions of their purpose.

The Code of Procedure for Measurement of Building Works (SMM7 Measurement Code) has been prepared for use with SMM7. The preface to the document sets out its objectives and it is noted that the application of the recommendations it contains is non-mandatory but is designed to represent good practice.

The Common arrangement of work sections for building works (CAWS) has been devised so that Specifications and Bills of Quantities are set out in a similar order. It is anticipated that a number of advantages should accrue from this approach. It should be easier to read documents relating to the same project together. It should also be easier to provide information to sub-contractors as the work sections have been carefully designed to assist in this respect. It is also hoped that the common arrangement will encourage greater consistency in the descriptions in Bills of Quantities and the content of Specifications. Each common arrangement work section contains a brief definition followed by lists of the items included or excluded. It will be necessary when using SMM7 to refer to the common arrangement to establish the work that should be measured under a particular work section.

Project Specification - a code of procedure for building works is the document that describes how specifications should be written in conformity with the common arrangement. Project specification is dealt with in more detail in Chapter 2.

Production Drawings - a code of procedure for building works is the document that describes how drawings should be prepared in conformity with the common arrangement.

SMM7 and the SMM7 Measurement Code have been published jointly by The Royal Institution of Chartered Surveyors (RICS) and The Building Employers Confederation (BEC). Members of the RICS and BEC sat on the Co-ordinating Committee for Project Information (CCPI) to ensure that their documentation was consistent in presentation with the BPIC documents. As previously stated the SMM7 Measurement Code is non-mandatory and this applies also to

the procedures, information and guidance given in the Common
Arrangement, Project Specification and Production Drawings codes
of procedure. The rules of SMM7 however are mandatory and will be
deemed to have been followed unless a clear reference is made in
the Bills of Quantities to any departures which have been made.

The CCPI have published a short guide to the use of the documents
published by the BPIC, RICS and BEC entitled "Co-ordinated Project
Information for building works, a guide with examples". This
guide details the problems which are encountered in the building
industry at the present time and sets down suggested guidelines to
try to ensure that these problems are minimised in the future.

Returning to SMM7 we set out below a brief comparison of the SMM7
Work Groups with the SMM6 trade headings. In Chapter 3 we have
compared SMM7 Work Sections with the SMM6 trade order in detail.

SMM6		SMM7	
A	General Rules		General Rules
B	Preliminaries	A	Preliminaries/General conditions
C	Demolition	C	Demolition/Alteration/Renovation
D	Excavation and Earthwork	D	Groundwork
		Q	Paving/Planting/Fencing/Site furniture
E	Piling and Diaphragm Walling	D	Groundwork
F	Concrete Work	E	In situ concrete/Large precast concrete
		F	Masonry
		H	Cladding/Covering
		K	Linings/Sheathing/Dry partitioning
		Q	Paving/Planting/Fencing/Site furniture
G	Brickwork and Blockwork	F	Masonry
H	Underpinning	D	Groundwork
J	Rubble Walling	F	Masonry
K	Masonry	F	Masonry
		H	Cladding/Covering

SMM6		SMM7	
L	Asphalt Work	J	Waterproofing
		M	Surface finishes
		Q	Paving/Planting/Fencing/Site furniture
M	Roofing	H	Cladding/Covering
		J	Waterproofing
N	Woodwork	G	Structural/Carcassing metal/timber
		H	Cladding/Covering
		K	Linings/Sheathing/Dry partitioning
		L	Windows/Doors/Stairs
		M	Surface finishes
		N	Furniture/Equipment
		P	Building fabric sundries
		Q	Paving/Planting/Fencing/Site furniture
P	Structural Steelwork	G	Structural/Carcassing metal/timber
Q	Metalwork	G	Structural/Carcassing metal/timber
		L	Windows/Doors/Stairs
		M	Surface finishes
		N	Furniture/Equipment
		Q	Paving/Planting/Fencing/Site furniture
		P	Building fabric sundries
R	Plumbing and Mechanical Engineering Installations	P	Building fabric sundries
		R	Disposal systems
		S	Piped supply systems
		T	Mechanical heating/Cooling/Refrigeration systems
		U	Ventilation/Air conditioning systems
		Y	Mechanical and electrical services measurement

SMM6		SMM7	
S	Electrical Installations	P	Building fabric sundries
		V	Electrical supply/power/ lighting systems
		W	Communications/Security/ Control systems
		Y	Mechanical and electrical services measurement
T	Floor, Wall and Ceiling Finishings	J	Waterproofing
		K	Linings/Sheathing/Dry partitioning
		M	Surface finishes
		Q	Paving/Planting/Fencing Site furniture
U	Glazing	H	Cladding/Covering
		L	Windows/Doors/Stairs
V	Painting and Decorating	M	Surface finishes
W	Drainage	R	Disposal systems
X	Fencing	Q	Paving/Planting/Fencing Site furniture

The following SMM7 Work Sections have no comparable SMM6 trade

- B Complete buildings
- X Transport systems
- – Additional rules – work to existing buildings

2 : Specification/Preambles

During recent years the preparation of a full project specification has been the exception rather than the rule in most sectors of the building industry when Bills of Quantities have formed part of the Contract documents. On most projects the annotation on Architect's drawings and the general descriptions of materials and workmanship section of Bills of Quantities have been the only form of specification given to a contractor. In the same period disputes and disagreements between parties to building contracts have become more common and the provision of a full project specification could have meant that the disagreement would not have arisen in the first instance.

The CCPI had as one of its objectives the improvement of the standard and quality of information given in specifications to provide a document more closely co-ordinated with other contract documents. This should enable contractors to more accurately price tender documents and eliminate the major proportion of disputes at the building stage caused by lack of information, ambiguity and incorrect or out of date specification.

The document published by the BPIC entitled "Project Specification - a code of procedure for building works" suggests guidelines to be followed when preparing a project specification. In the same manner as other documents published by the BPIC the guidance on coverage within "Project Specification" is in common arrangement work section order. It is intended that it should be used in conjunction with the production drawings code so that the two documents compliment each other and the drawings and specifications thus prepared cover all the drawn and descriptive information needed for tender documents and later to construct the building.

The code recommends that the responsibility for preparing the project specification rests with the designer as it is the designer who should co-ordinate the necessary information regarding materials, workmanship requirements, working conditions and the like for a project. Engineers, Quantity Surveyors and other specialists will often contribute to the specification; Engineers with specifications for their own work and Quantity Surveyors during the preparation of the Bills of Quantities with comments, clarifications and additions.

Specifications when prepared have been presented in many different formats in the past. The work carried out by the CCPI is intended to result in more uniformity of presentation. It is recommended that the arrangement of the specification shall closely follow the Common arrangement of work sections for building works. Within a work section the arrangement will vary and be dependent upon the type of specification to be included and how the specification best relates to the drawings and measured work to provide ease of cross-reference. The old established method of separately specifying materials and workmanship may still be used if necessary but in practice a combined clause may more suitably describe the work and simplify cross-referencing. It is recommended that both Bills of Quantities items and annotations on drawings should contain cross-references to the specification. If this procedure is followed it will enable descriptions to be very much shorter than at present.

In the same way that many Quantity Surveyors have been using standard libraries of descriptions for the preparation of Bills of Quantities the code recommends that specifications would be more easily and accurately prepared using standard libraries of specification. Use of standard libraries of specification will have the advantages of providing the designer with a check-list of information to be provided and of enabling all members of the pre-contract team to have an early copy. This will mean that Quantity Surveyors will be able to follow the guidelines and commence measurement before the final specification is produced as the designer will be able to liaise with them with regard to omissions from and additions to the standard and the agreement of specification references. Failure to provide an early copy of the specification could result in delays to the procurement programme. The code suggests that a separate document could make it easier to relate the specification to the measured work. It will still be necessary for the Bills of Quantities to contain additional information which may have an effect on prices.

Examples of suggested specification formats are given in "Project Specification - a code of procedure for building works" published by the BPIC and also in the guide entitled "Co-ordinated project information for building works, a guide with examples" published by the CCPI.

3 : SMM6 to SMM7
– A Detailed Comparison

This chapter has been written to provide the practitioner, who has a knowledge of the rules of SMM6, with an easy route to the changes in measurement rules required by SMM7.

The main part of the chapter is tabulated in SMM6 trade and item order with the corresponding SMM7 rules adjacent. Where the words "no change" "unchanged" or the like are used herein they are meant to infer that the general requirement of the rule(s) has remained the same as SMM6. These words do not however mean that the wording of the rule(s) is identical to that in SMM6 and the particular rules should be read to establish the exact requirements of SMM7.

Certain of the work sections in SMM7 have no corresponding rules in SMM6; these rules have been collated immediately following this introduction. To assist in finding a particular SMM7 Work Section a list of Work Sections in alpha/numeric order with page numbers has been included as Appendix A at the end of this book.

The following work sections/items were not previously specifically covered by SMM6:

    H11:   **Curtain walling**
              - Curtain walling is defined as non load bearing walls of wood or metal framing fixed as an intergrated assembly complete with windows, and opening lights, glazing and infill panels.

It is to be measured in square metres stating whether flat, sloping or curved. Extra over items of infill panels, perimeter angle and closers are to be measured in linear metres and opening lights and doors are to be enumerated (except doors supplied with the unit). All items supplied with the unit are deemed to be included.

H14: **Concrete rooflights/pavement lights**

- Concrete rooflights/pavement lights, also covers vertical units (other rooflights are measured under Section L.12).

Rooflights and pavement lights are to be measured in square metres stating the number and vertical units are to be enumerated with a detailed description.

K41: **Raised access floors**

- Raised access floors are defined as false platforms of dry construction raised above the structural floor slab to create space for the distribution of services. The construction normally comprises a height adjustment. The thickness of the panel and the height of the cavity are to be stated.

P11: **Foamed/Fibre/Bead cavity wall insulation**

- Foamed/Fibre/Bead cavity wall insulation is to be measured in square metres of area actually filled stating the thickness, type, quality, method of application and associated works.

P22:   **Sealant joints**

- Sealant joints are joints which for special reasons cannot reasonably be included in another Work Section.

R20:   **Sewage pumping**   ) To be measured
) in accordance
R21:   **Sewage treatment/**   ) with Work
        sterilisation   ) Section Y

X10:)
to   )   **Transport systems**
X32:)

- Transport systems include lifts, escalators, moving pavements, hoists, cranes, travelling cradles, goods distribution/mechanised warehousing, mechanical document conveying, pneumatic document conveying and automatic document filing and retrieval. Items are to be enumerated giving full specification details, and information regarding builder's work, identification, testing and commissioning, temporary operation, preparing drawings and operating manuals.

-   **Additional rules - work to existing buildings**
    - the additional rules sections comprise items for bonding and jointing new to existing, stripping off, removing, taking down, making good, cutting and cutting holes in connection with Work Sections H, J, K, L, M and Mechanical and Electrical Services

All other Work Sections are referred to in the tabulation which follows.

SMM6		SMM7	
Clause	Heading	Clause	Heading/Comment
A	**GENERAL RULES**	-	**GENERAL RULES**
A.1	Introduction	1.1 & 1.2	No change
A.2	Bills of Quantities		
A.2.1	Bills of Quantities	1.1 and 10.1-2	No change except:- 1) the term "provisional" quantity is no longer used; these quantities are to be identified as "approximate" 2) provisional sums shall be identified as either for defined or undefined work. The definition of "defined work" is set out in general rule 10.3
A.2.2	Rules of measurement for work not covered	11.1	Now states such rules shall as far as possible conform with those given for similar work
A.3	Measurement		
A.3.1	Term metre	12.1	Now covered under symbols and abbreviations
A.3.2	Work measured net as fixed	3.1	No change except where otherwise stated in a measurement rule applicable to the work

14

SMM6		SMM7	
Clause	Heading	Clause	Heading/Comment
A.3.2	Measurement taken to nearest 10mm	3.2	Worded differently but no change in meaning
A.3.3	Deductions of voids	3.4	Unchanged from SMM6 unless otherwise stated in rules within specific Work Sections
A.3.4	Classification between two limiting dimensions	4.4	No change - See also clause 12.1 symbols for exceeding and not exceeding
A.4	Descriptions		
A.4.1	Order of stating dimensions	4.1	No change
A.4.2	Items deemed to be included	4.6(a)-(g)	As SMM6 with the addition of "Assembling" materials and goods
A.4.3	Junctions between curved and straight work	-	Now covered where applicable in coverage rules for each Work Section
A.4.4 & 5	Labours given in items on which they occur		No coverage. Many labours are now deemed to be included. Measurable labours are detailed in the various Work Sections
A.5	**Drawn Information**		
A.5.1.a	Location drawings	5.1	Little change except that plans sections and elevations to be provided in lieu of a general locating drawing
A.5.1.b	Component details	5.2	Now worded "Component drawings"

SMM6		SMM7	
Clause	Heading	Clause	Heading/Comment
A.5.1.c	Bill diagram	5.3	Now worded "Dimensioned diagrams" and may be used in place of a dimensioned description but not in place of an item otherwise required to be measured
A.5.2	Requirements for detailed description deemed to have been complied with if drawn information provided and such information indicates fully the items to be described	5.3	As immediately above
A.6	Standard Products	6.1	Worded differently but no change in meaning
A.7	Quantities		
A.7.1-2	Unit of billing	3.3	No change
A.7.3	Elimination of an item	-	Not mentioned. We consider that if this situation should arise the provisions of SMM6.A.7.3 should be applied and the item should be fully described and enumerated unless it is an insignificant item in relation to cost
A.8	Provisional and prime cost sums		
A.8.1.a	Provisional sums	10.2-6	Now to be identified in the following two categories:-

SMM6		SMM7	
Clause	Heading	Clause	Heading/Comment
			1) "Defined Work" Clauses 10.3 & 4 - Work not completely designed but for which the following information shall be provided
			a) the nature and construction of the work
			b) a statement of how and where the work is fixed to the building and what other work is to be fixed thereto
			c) a quantity or quantities which indicate the scope and extent of the work
			d) any specific limitations and the like
			Due allowance for "Defined Work" shall be deemed to have been made in programming planning and pricing of Preliminaries for a) to b) above except where a variation in other work measured in detail would give rise to an adjustment
			2) "Undefined Work" Clauses 10.5 & 6 - Work where the information required in accordance with

SMM6		SMM7	
Clause	Heading	Clause	Heading/Comment
			a) to d) above cannot be given. Due allowance (for "Undefined Work") shall be deemed not to have been made in programming planning and pricing Preliminaries
A.8.1.b	Prime cost sums	A51	Dealt with in Preliminaries
		A53	Work by Statutory Authorities or public companies carrying out statutory work shall now be covered by a provisional sum
A.9	Work in special conditons	7	Work of special types
A.9.1	Alterations and work in existing buildings	7.1(a) 13.1-3 and Additional Rules	Now defined as work on, in or under work existing before the current project. The "Additional Rules" for work in existing buildings are to be read in conjunction with the rules to the appropriate section. A description of the additional Preliminaries and general conditions pertinent to the work shall be given
A.9.2	Work carried out in or under water	7.1(d)	No change
A.9.3	Work carried out in compressed air	7.1(e)	No change

SMM6		SMM7	
Clause	Heading	Clause	Heading/Comment
			Additional categories
		7.1.(b)	Work to be carried out and subsequently removed (other than temporary work)
		7.1.(c)	Work outside curtilage of site

SMM6		SMM7	
Clause	Heading	Clause	Heading/Comment
B	**PRELIMINARIES** **Preliminary particulars**	A	**PRELIMINARIES/ GENERAL CONDITIONS**
B.1	Project, parties and consultants	A10	Project Particulars
		A13	Essentially no difference except "Description of the Work" specific requirements to be stated
B.2	Description of site	A12	The site/Existing Buildings
			Essentially no difference except "means of access" not required to be given with this item - see A35
B.3	Drawings and other documents	A11	Drawings - location, assembly and component drawings will be given in accordance with the Common Arrangement and Code for Production Drawings
	Contract		
B.4	Form, type and conditions of contract	A20	The Contract/Sub Contract Employer's insurance responsibility and Performance guarantee bond added under this heading
B.5	Contractor's liability	-	Now dealt with more simply under A20 above - assumes contract will cover
B.6	Employer's liability	A20:1.4	Employer's insurance responsiblity

SMM6		SMM7	
Clause	Heading	Clause	Heading/Comment
B.7	Local Authorities fees and charges	A:C1	Rates, fees and charges are deemed to be included for works of a temporary nature
B.8	Obligations and restrictions imposed by the Employer		Fixed charges and time related charges apply to all preliminary items below
B.8.1a	Access to and possession or use of the site	A34:1.2 A35:1.3 A35:1.4	Maintain adjoining buildings Access Use of the site
B.8.1b	Limitations of working space	−	Not specifically mentioned − covered by A35:1.1, A35:1.2, A35:1.8
B.8.1c	Limitations of working hours	A35:1.6 A35:1.7	Start of work Working hours
B.8.1d	The use or disposal of any materials found on site	A35:1.5	Use or disposal of materials found
B.8.1e	Hoarding, fences, screens, temporary roofs, temporary nameboards and advertising rights	A36:1.3 A36:1.4 −	Temporary fences, hoardings, screens and roofs Nameboards Advertising rights are not specifically mentioned
B.8.1f	Maintenance of existing live drains or other services	A34:1.4	Maintain live services
B.8.1g	The execution or completion of the work in any specific order or in sections or phases	A35	Employer's requirements or limitations, details stated (with appropriate sub-headings)

SMM6		SMM7	
Clause	Heading	Clause	Heading/Comment
B.8.1h	Maintenance of specific temperatures and humidity levels	A36:1.6	Temperature and humidity. No alternative given for provisional or prime cost sums
B.8.1j	Temporary accommodation and facilities for the use of the employer	A36:1.1 A36:1.2 A36:C2	Offices Sanitary accommodation. Heating, lighting. Cleaning and maintenance are deemed to be included. Furnishing and attendance still need to be stated
B.8.1k	Installation of telephones for the use of the Employer and the cost of his calls to be given as a provisional sum	A36:1.7 A36:1.9	Telephone/Facsimile installation and rental/maintenance Telephone/Facsimile call charges - Provisional Sum
B.8.1l	Any other obligation or restriction	A36:1.5 A34:1.7 A35:1.8 A36:1.8	Technical and surveying equipment - Now an SMM heading if Employer requires Contractor to provide))Others)
	Works by Nominated Sub-contractors, goods and materials from nominated suppliers and works by public bodies		
B.9	Works by nominated sub-contractors	A51	Nominated sub-contractors - the work is to be described in accordance with the rules for defined provisional sums

22

SMM6		SMM7	
Clause	Heading	Clause	Heading/Comment
B.9.2	General attendance	A42:1.16	General attendance on nominated sub-contractors - intended to be a single item for all sub-contracts (See coverage rule C3 for items deemed to be included)
B.9.3	Other attendance	A51:1.3 and D8	Special attendance now includes standing scaffolding required to be altered or retained. The Code of Procedure for measurement elaborates on the detail to be given
B.10	Goods and materials from nominated suppliers	A52	Nominated suppliers No reference is made of the items which are to be included in the pricing of 'Fix only' work. It would seem that a coverage rule should have been included stating that "Unloading, hoisting etc." is deemed to be included as SMM6 clause B.10.2
B.11	Works by Public Bodies	A53	Work by statutory authorities (See definition rule D10)
B.12	Works by others directly engaged by the Employer	A50	Work/Materials by the Employer. This clause is extended to cover the provision of materials only by the Employer. Generally the above provisions are essentially unchanged but are repositioned

SMM6		SMM7	
Clause	Heading	Clause	Heading/Comment
B.13.1a	General facilities and obligations. Plant, tools and vehicles	A42:1.14 A43 A43:1.1 A43:1.2 A43:1.3 A43:1.4 A43:1.5 A43:1.6 A43:1.7 A43:1.8 A43:1.9	Small plant and tools Mechanical plant Cranes Hoists Personnel transport Transport Earthmoving plant Concrete plant Piling plant Paving and surfacing plant Others, details stated and statement in definition rules that items listed not exhaustive. Plant no longer required to be itemised in each work section.
B.13.1b	Scaffolding	A44:1.3 A44:1.4	Access scaffolding Support scaffolding and propping
B.13.1c	Site administration and security	A40 A42:1.12 A34:1.5	Management and staff Security (under Services and facilities) Security (under Employers Requirements)
B.13.1d	Transport for workpeople	A43:1.3	Personnel transport
B.13.1e	Protecting the Works from inclement weather	A34:1.6 and A42:1.11	Protection of work in all sections - intended to be a much broader item as protection related to each work section is no longer required to be given at the end of each section
B.13.1f	Water for the Works	A42:1.4	Water
B.13.1g	Lighting and power for the Works	A42:1.1 A42:1.2	Power Lighting

SMM6		SMM7	
Clause	Heading	Clause	Heading/Comment
B.13.1h	Temporary roads, hardstandings, crossings and similar items	A44:1.1 A44:1.2 A44:1.6	Temporary roads Temporary walkways Hardstanding
B.13.1j	Temporary accommodation for the use of the Contractor	A41	Site accommodation - D4 defines the extent deemed to be covered
B.13.1k	Temporary telephones for the use of the Contractor	A42:1.5	Telephone and administration
B.13.1l	Traffic regulations	A44:1.7	Traffic regulations
B.13.1m	Safety, health and welfare of workpeople	A42:1.6	Safety, health and welfare
B.13.1n	Disbursements arising from the employment of workpeople	—	Not specifically mentioned but see definitions of fixed and time-related charges which will cover as appropriate
B.13.1p	Maintenance of public and private roads	A34:1.3 A42:1.13	Maintain public and private roads (Employer's requirements) Maintain public and private roads (Services and facilities)
B.13.1q	Removing rubbish, protective casings and coverings and cleaning the works on completion	A42:1.8 A42:1.11 A42:1.9	Rubbish disposal Protection of work in all sections Cleaning
B.13.1r	Drying the works	A42:1.10	Drying out

SMM6		SMM7	
Clause	Heading	Clause	Heading/Comment
B.13.1s	Temporary fencing, hoardings, screens, fans, planked footways, guardrails, gantries and similar items	A44:1.5	Hoardings, fans, fencing etc.
B.13.1t	Control of noise, pollution and all other statutory obligations	A34:1.1	Noise and pollution control
B.14	Contingencies		No specific item for contingencies but:-
		A54	Provisional Work and reference to defined and undefined provisional sums as general rule 10
		A55	Dayworks - labour, materials and plant
There are no comparable sections in SMM6 for the following:-			
		A30	Employers requirements: Tendering/ sub-letting/supply
		A31	Employers requirements: Provision, content and use of documents
		A32	Employers requirements: Management of the Works
		A33	Employers requirements: Quality Standards/ Control
		A37	Employers requirements: Operation/ Maintenance of the finished building
		A42	Contractor's general cost items: Services and facilities

SMM6		SMM7	
Clause	Heading	Clause	Heading/Comment
		A42:1.3	Fuels
		A42:1.7	Storage of materials
			General Comment
			Although much change in order of preliminary items actual requirements relating to individual items vary little except for the amount of information that should be supplied describing the works and the requirement for fixed and time-related charges against many items. It is suggested that the requirements relating to fixed and time-related changes can be satisfied by the appropriate provision of headed columns in bills of quantities

SMM6		SMM7	
Clause	Heading	Clause	Heading/Comment
C	**DEMOLITION**	C	**DEMOLITION/ALTERATION/ RENOVATION**
		C10	**Demolishing Structure**
	Generally	C30	**Shoring**
			Note:- References below to C10 apply equally to C30
C.1	Information	C10:P1	Information
			A description of the work is no longer an alternative. Location drawings have to accompany the Bills of Quantities
	Specific methods	C10:S1	Specific means - see also C20:S1
C.2	Plant	A43	Dealt with in preliminaries
C.3	Demolition		
C.3.1	Location of demolitions	C10:P1	Location and extent of existing structures to be demolished is to be shown on the location drawings - see also C20:D2
C.3.2	Old materials become the property of the Contractor	C10:D1	Materials arising from demolitions are the property of the Contractor unless otherwise stated - see also C20:D1
	Credits	-	No provision. Surveyors will have to make their own assessment and treat each job on it merits

SMM6		SMM7	
Clause	Heading	Clause	Heading/Comment
	Clearing away deemed to be included	C10:C1(a)	Disposal of materials other than those remaining the property of the Employer or those for re-use deemed to be included - see also C20:C2(a)
C.3.3	Old materials to remain the property of the Employer	C10:1-3.1.1.1	Materials remaining the property of the Employer - see also C10:S2 and C20:S2 for setting aside and storing
C.3.4	Old materials re-used	C10:1-3.1.1.2	Materials for re-use No longer a reference to non-adjustment of measured quantities of new work in which old materials are re-used
C.3.5	Method of disposal - restrictions imposed by the Employer	C10:S3	Employers restrictions on method of disposal of materials - see also C20:S3
C.3.6	Handling and disposal of toxic or other dangerous materials	C10:1-3.1.1.6	Toxic or other special waste - see also C20
C.4	Measurement		Measurement generally unchanged but many items reworded
C.4.1	Demolishing individual structures (or parts)	C10:2 C10:3	Demolishing individual structures Demolishing parts of structures
	Clearing site of all structures	C10:1	Demolishing all structures
	Level to which to be demolished to be stated	C10:1-3.1.1.*	Levels to which structures are demolished

SMM6		SMM7	
Clause	Heading	Clause	Heading/Comment
	Making good remaining structures and finishings	C10:1-3.1.1.3	Making good structures
	Leaving parts of walls temporarily in position as buttresses	C10:1-3.1.1.4	Leaving parts of existing walls temporarily in position to act as buttresses
		C10:1-3.1.1.5	Temporarily diverting, maintaining or sealing off existing services - not previously mentioned in SMM6 although would have been included in description or itemised separately
		C20	**Alterations - spot items**
C.4.2	Cutting openings or recesses:-	C20:5	Cutting openings or recesses
	Size of opening, type and thickness of structure	C20:5.2	Dimensioned description sufficient to identify type and thickness of existing structure
	Treatment around opening or recess	C20:5.2.0.3	Inserting new work - details stated see M3 and D3 also C2 all new fixing and jointing materials deemed to be included
	Making good the structure deemed to be included	C20:5.2.0.1	Making good structure to be stated
	Extending finishings to be included in descriptions	C20:5.1.0.2	Extending and making good finishings

SMM6		SMM7	
Clause	Heading	Clause	Heading/Comment
C.4.3	Cutting back projections	C20:6	Cutting back projections
	Cutting to reduce thickness	C20:7	Cutting to reduce thickness
C.4.4	Removal of fittings and fixtures prior to demolition	C20:1	Removing fittings and fixtures
	Those set aside to be stated	C20:S2	Setting aside and storing materials remaining the property of the Employer or those for reuse
	Making good structure and finishings to be described	C20:1.1.0.1 C20:1.1.0.2	Making good structures. Extending and making good finishings
C.4.5	Removing engineering and plumbing installations	C20:2	Removing plumbing and engineering installations - dimensioned descriptions sufficient for identification including details of making good structure and finishings and disposal of toxic waste i.e. asbestos lagging
C.4.6	Removing finishes or coverings stating nature and quantity	C20:3 C20:4	Removing finishings Removing coverings Dimensioned descriptions sufficient for identification required for both
C.4.7	Temporary screens and temporary roofs	C10:7 C20:10 C10:6 C20:9	Temporary screens Temporary roofs

31

SMM6		SMM7	
Clause	Heading	Clause	Heading/Comment
	Clearing away deemed to be included	C10:6-7.1.0.4 C20:9-10.1.0.4	Clearing away to be stated
	Can be measured in square metres or enumerated	C10:6-7.1.0.* C20:9-10.1.0.* C10:6-7.1.0.2-3 C20:6-7.1.0.2-3	Dimensioned description to be given. Maintaining duration and adapting details to be stated
	Weatherproof and dustproof screens to be so described	C10:S4 C20:S5	Details of weather and dust proofing requirements to be given as supplementary information
C.5.1	Shoring and scaffolding incidental to demolition of individual structures (or parts thereof) is deemed to be included	C30:C1(b)	Temporary support is deemed to be included
	Shoring and scaffolding incidental to cutting openings or recesses are deemed to be included	C20:C1	Shoring or scaffoldinng to the work are deemed to be included
C.5.2	Shoring (other than incidental to demolitions)	C30:4	Support of structures not be demolished
			Maintaining the shoring; duration now to be stated
		C30:5	Additional rules included for support of roads and the like

SMM6		SMM7	
Clause	Heading	Clause	Heading/Comment
	Protection		
C.6	Protecting the work	A34:1.6 A42:1.11	Dealt with in Preliminaries under "Employer's requirements" and "Contractor's general cost items"
		General Note	
		Refer also to "Additional rules - work to existing buildings"	

SMM6		SMM7	
Clause	Heading	Clause	Heading/Comment
There are no comparable sections in SMM6 for the following:-			
		C40	**Repairing/Renovating concrete/brick/block/ stone**
		C40:1	Cutting and defective concrete and replacing with new
		C40:2	Resin or cement impregnation/injection
		C40:3	Cutting and decayed, defective and cracked work
		C40:4	Repointing
		C40:5	Removing stains and the like
		C40:6	Cleaning surfaces
		C40:7	Inserting new wall ties
		C40:8	Re-dressing to new profile
		C40:9	Artificial weathering
		C41	**Chemical dpc's to existing walls**
		C41:1	Chemical damp-proof courses .1 to brickwork .2 to blockwork .3 to stonework

SMM6		SMM7	
Clause	Heading	Clause	Heading/Comment
		C50	**Repairing/Renovating metal**
		C51	**Repairing/Renovating timber**
		C52	**Fungus/Beetle eradication**
			References to C50: apply equally to C51 and C52
		C50:1	Repairing metal
		C50:2	Repairing timber
		C50:3	Treating existing timber

35

SMM6		SMM7	
Clause	Heading	Clause	Heading/Comment
D	**EXCAVATION AND EARTHWORK**	D	**GROUND WORK**
	Generally	D20	**Excavating and filling**
		Q20	**Hardcore/Granular/ Cement bound bases/ sub-bases to roads/ pavings**
			Note:- Reference below to D20 apply equally to Q20 where where applicable
D.1	Excavation in underpinning and drainage dealt with in Sections H & W	-	Underpinning is dealt with under work Section D50.
			Drainage is dealt with under work sections R12 and R13 although excavation for manholes and certain details for trenches are referred back to this work Section
D.2	Information	D20:P1 (a-d)	Ground water level and date to be stated Details of trial pits or bore holes including their location
		(e)	Features retained
		(f)	Live over or underground services including their location Pile sizes and layout in accordance with Sections D30-D32 where applicable
D.3	Soil description	-	Generally dealt with under Clause P1
D.4	Plant	A43	Dealt with in preliminaries

SMM6		SMM7	
Clause	Heading	Clause	Heading/Comment
	Site preparation		
D.5	Removing trees	D20:1.1	Removing trees and stumps: girth ranges
		D20:1.2	girth 600mm - 1.50m girth 1.50 - 3.00m girth > 3.00m and stating the girth
			Small trees and stumps (trees not exceeding 600mm girth) are removed within the items of "Clearing site vegetation"
		D20:C1	Grubbing up roots and filling voids now deemed to be included
D.6	Removing hedges	D20:1.3 & D1	Included in item of "Clearing site vegetation"
D.7	Clearing undergrowth	D20:1.3	Now described as "Clearing site vegetation" and includes for removing bushes, scrub, undergrowth, hedges and trees and tree stumps ≤ 600mm girth
D.8	Lifting turf	D20:1.4	No requirement for disposal to be stated
D.9	Topsoil	D20:2.1	No change
	Excavation		
D.10	Volume to be measured	D20:M3	No change except the additional wording that "no allowance is made .... for extra space for working space"

37

SMM6		SMM7	
Clause	Heading	Clause	Heading/Comment
D.11	Depth Classification	D20:2.*. 1-4	No change except that that the "not exceeding 4.00m range" no longer appears as it is covered by the words "and thereafter in 2.00m stages"
D.12	Working space	D20:6	Now only required where the face of excavation is < 600mm from the face of formwork, rendering, tanking or protective walls - irrespective of depth and is measured in square metres by multiplying the girth of the formwork, etc. by the depth of excavation. Working space to reduce levels and basements is now grouped together as a single item. Working space for post tensioned concrete is no longer mentioned
D.13	Types of excavation	D20:2.*.*.1	Commencing level only stated where exceeding 0.25m below existing ground level.
		D20:2.2-8	No requirements to state curved trenches. The extra cost of curved excavation is deemed to be included with curved earthwork support (C3). Trenches ≤ 0.30m now to be measured in cubic metres.
		D20:2.8.*.*	A new item of "Excavation; to bench sloping ground to receive filling ..." has been included. The Code of Procedure for Measurement explains that this item

38

SMM6		SMM7	
Clause	Heading	Clause	Heading/Comment
			covers "the excavation of steps in the face of a slope to prevent the slippage of subsequent filling"
D.13.8	Excavating trenches to receive service pipes	P30:1	Size of service to be given as ≤ 200mm or where > 200mm the actual size. Deemed to include earthwork support, consolidation of trench bottom, trimming excavations, special protection of services, backfilling with and compaction of excavated materials and disposal of surplus excavated materials.
D.13.9 & 10	Excavating alongside and around services crossing excavation	D20:3.2 & 3	No change
D.13.11 & 12	Breaking up existing materials	D20:4 & 5. 1-5.* and D20:C2	No change except that breaking out in working space is deemed to be included
D.13.13	Excavating below ground water level	D20:3.1 and D20:C2	Excavating below ground water level in working space is now deemed to be included
D.14-24	**Earthwork support**	D20:7 D20:C3	Generally as SMM6 but:- Curved earthwork support is deemed to include the extra cost of curved excavation

SMM6		SMM7	
Clause	Heading	Clause	Heading/Comment
D.18	Trenches etc. below faces of excavation	-	No provision
D.22	Unstable ground	D20:D8	The definition of unstable ground has been extended to include loose gravel and the Code of Procedure for Measurement extends this further by suggesting that strata could be said to be unstable when "the newly excavated face will not remain unsupported sufficiently long enough to allow the necessary support to be inserted"
	Disposal of Water		
D.25	Surface water	D20:8.1 & 2	No change
D.26	Water in the ground		
D.27-32	**Disposal of Excavated material**	D20.8.3.1&2 and D20:C4	No requirement now to distinguish between top soil and other excavated materials
D.33-37	Filling	D20:9 10 & 11	All filling is now classified as over 250mm or not exceeding 250mm thick and given in cubic metres. Filling to make up levels no longer to be given in square metres stating average depth
		D20:11	New classification of filling to external planters included - the position to be given when not at ground level

SMM6		SMM7	
Clause	Heading	Clause	Heading/Comment
D.38 & 39	Hand-packing	D20:12	Surface packing to vertical or battered surfaces now given in square metres irrespective of width
			No reference now made to sinkings
			Definition of battered surfaces now included i.e. > 15 degrees from horizontal
D.40-43	Surface treatments, Compaction, Trimming and blinding	D20:13	Generally unchanged except no requirement to identify trimming to curved work
		D20:13.1	Classification now included for the application of herbicides
		D20:13.5	Classification also included for preparing subsoil for top soil
D.44	**Landscaping**	Q30	**Seeding/turfing**
		Q31	**Planting**
		Q30:1 & Q31:1	Cultivating
		Q30:2 & Q31:2	Surface applications
		Q30:3	Seeding
		Q30:4	Turfing
		Q30:5	Turfing edges of seeded areas
		Q30:6 & Q31:10	Protection

SMM6		SMM7	
Clause	Heading	Clause	Heading/Comment
		Q31:3	Trees
		Q31:4	Young nursery stock trees
		Q31:5	Shrubs
		Q31:6	Hedge plants
		Q31:7	Herbaceous plants
		Q31:8	Bulbs, corms and tubers
		Q31:9	Mulching after planting
			All of the above items would have been covered in general terms by SMM6 clause D44 but the rules for the particular works are now given in detail
	Protection		
D.45	Protecting the work		Dealt with in preliminaries under "Employer's requirements" and "Contractor's general cost items"
		A34:1.6 and A42:1.11	

There was no specific section for the following item in SMM6:-

		Q23	**Gravel/Hoggin roads/ pavings**
	This type of construction would have been measured under SMM6 clause D.33		To be measured in square metres of "Roads" or "Pavings" stating the thickness

42

SMM6		SMM7	
Clause	Heading	Clause	Heading/Comment
E	**PILING AND DIAPHRAGM WALLING**	D	**GROUNDWORK**
			The various types of piling have now been given individual Work Sections within the Groundwork Work Group. To enable easy comparison of each type of piling they have been separately compared with SMM6 and follow each other in Work Section sequence herein followed by Diaphragm Walling
			The Work Sections compared within this section are as follows:-
		D30	**Cast in place concrete piling**
		D31	**Preformed concrete piling**
		D32	**Steel piling**
		D40	**Diaphragm walling**

SMM6		SMM7	
Clause	Heading	Clause	Heading/Comment
	Piling	D30	**Cast in place concrete piling**
E.1	Information	D30:P1 (a)-(c)	No change
		(d)	The relationship to adjacent buildings to be stated
E.2.1-2	Soil description	D30:P2 (a) & (b)	No change
E.2.3		D30:P3(a)	No change
E.3	Plant	A43	Dealt with in preliminaries
E.4	Piling description		
E.4.1.a	Bored cast-in-place concrete piles	D30:1	Bored piles
E.4.1.b	Driven cast-in-place concrete piles	D30:2	Driven shell piles as definition rules D1 and D2
E.4.1.c	Pre-formed concrete piles		Dealt with in Section D31
E.4.1.d	Pre-formed prestressed concrete piles		Dealt with in Section D31
E.4.1.e	Pre-formed concrete sheet piles		Dealt with in Section D31
E.4.1.f	Timber piles		Not mentioned
E.4.1.g	Isolated steel piles		Dealt with in Section D32
E.4.1.h	Interlocking steel piles		Dealt with in Section D32
E.4.1.j	Other piles		No provision
-	Preliminary and test piles	D30:1.*.*.1	No change

SMM6		SMM7	
Clause	Heading	Clause	Heading/Comment
–	Contiguous bored piles	D30:1.*.*.2	No change
	raking piles	D30:1.*.*.3	Raking inclination ratio stated – no longer required to be stated in 10 degree increments
E.4.2	Above classifications to be grouped under sub-headings stating:-		No specific mention of grouping under sub-headings but this would still apply
	nominal cross sectional size	D30:1.1	Nominal diameter stated
	materials	D30:S1-4	Kind and quality of materials and MIX details and tests
E.4.3	For piles other than cast-in-place concrete and interlocking steel piles		Dealt with in Section D31
E.4.4	Cast-in-place concrete piles	D30:1 D30:2	Bored piles Driven shell piles
E.4.4.a	The total number of piles	D30:1.1.1	1. Total number, commencing surface stated
E.4.4.b	The total completed length of the piles given in metres	D30:1.1.2	2. Total concreted length
E.4.4.c	The total driven or bored depth	D30:1.1.3	Total length maximum depth stated – no longer given in ranges
E.4.5. a-c	Interlocking steel piles		Dealt with in Section D32
E.4.6.a	Boring through rock	D30:5.1	Breaking through obstructions – now to be given in hours

SMM6		SMM7	
Clause	Heading	Clause	Heading/Comment
E.4.6.b	Permanent casing	D30:6.1 D30:6.2	To be given in metres in lengths $\leq$ 13m and $>$ 13m and stating the number of casings. The internal diameter and wall thickness is also to be given. Permanent casings are not now measured as extra over the piling
E.4.6.c	Placing concrete by tremie pipe		Not mentioned
E.4.7.a	Pre-boring where specifically required	D30:3 D30:M2 D30:C2	Maximum depth to be stated Measured only when specifically required Pre-boring is deemed to include grouting up voids between sides of piles and bores
E.4.7.b	Backfilling empty bores	D30:4	No change
E.4.7.c	Jetting		Dealt with in Section D31
E.4.7.d	Filing hollow piles		Dealt with in Section D31
E.4.7.e	Pile extensions		Dealt with in Sections D31 and D32
E.4.7.f	Cutting interlocking piles to remove tops		Dealt with in Section D32
E.4.8.a	Cutting off tops of isolated piles	D30:7.1.1.1 D30:15	Now to be given in metres stating number nominal diameter and details of permanent casings. Now deemed to include preparation and integration of reinforcement into pile cap or ground

SMM6		SMM7	
Clause	Heading	Clause	Heading/Comment
			beam and disposal. Code of Procedure for Measurement states that the responsibility as between main contractor and sub-contractor for cutting off tops of piles and for disposal of materials should be made clear in the Bills of Quantities. Cutting off tops of test piles should be given separately.
E.4.8.b	Preparing concrete pile heads and reinforcement for capping	D30:C5	Now deemed to be included in cutting off tops of piles
E.4.8.c	Forming enlarged bases	D30:5.2 & 3	Now measured as extra over piling
E.4.8.d	Heads and shoes	D30:C4	Permanent casings are deemed to include driving heads and shoes
E.4.8.e	Removal of piles		Not mentioned
E.4.8.f	Cutting interlocking steel piles		Dealt with in Section D32
E.5	Reinforcement	D30:8.1	Nominal size of bars stated
		D30:8.2.1	Nominal size of bars and diameter of piles to be stated for helical bars.
		D30:C6	Tying wire spacers links and binders which are at the discretion of the Contractor are deemed to be included.
		D30:S7	Kind and quality of materials to be given

SMM6		SMM7	
Clause	Heading	Clause	Heading/Comment
E.6	Disposal	D30:9	Now required to state whether disposed of on or off site. Specified locations and handling details to be stated.
		D30:M6	Method of calculating Volume defined. The volume of enlarged bases is to be added
E.7	Delays		
E.7.1	Standing time for rigs	D30:10.1 D30:M7 D30:C7	No change
E.7.2	Boring through artificial obstructions	D30:5.1	Now given as extra over piling
E.8	Tests	D20:11	Details of testing to be stated. No specific tests are now mentioned
E.9	Quantities		Generally dealt with under measurement and coverage rules as defined in the foregoing comments and as follows:-
E.9.1.a	Driven or bored length	D30:M1	Measured along axes of the piles from commencing surface to bottom of shafts or bottom of casings
E.9.1.b	Concreted lengths	D30:M1 D30:C1	As above and work is deemed to include concrete placed in excess of the completed length
E.9.1.c	Lengths for preformed piles		Dealt with in Section D31
E.9.1.d	Area of interlocking piles		Dealt with in Section D32

SMM6		SMM7	
Clause	Heading	Clause	Heading/Comment
E.9.1.e	Pile extensions		Not applicable
E.9.1.f	Volume of surplus excavated material	D30:M6	Volume calculated from the nominal cross-sectional size of piles and their lengths measured as D30:1 & 2.2.1.2.*. Volume of enlarged bases to be added

SMM6		SMM7	
Clause	Heading	Clause	Heading/Comment
	Piling	D31	**Pre-formed concrete piling**
E.1	Information	D31:P1(a)-(c)	No change
		(d)	The relationship to adjacent buildings to be stated
E.2.1-2	Soil description	D31:P1(a) & (b)	No change
E.2.3.		D31:P3(a)	No change
E.3	Plant	A43	Dealt with in preliminaries
E.4	Piling description		
E.4.1.a & b	Bored and driven cast-in place concrete piles		Dealt with in Section D30
E.4.1.c	Pre-formed concrete piles	D31:1 D31:4	Reinforced piles Hollow section piles
E.4.1.d	Pre-formed prestressed concrete piles	D31:2	Prestressed piles
E.4.1.e	Pre-formed concrete sheet piles	D31:3	Reinforced sheet piles
E.4.1.f	Timber piles		Not mentioned
E.4.1.g	Isolated steel piles		Dealt with in Section D32
E.4.1.h	Interlocking steel piles		Dealt with in Section D32
E.4.1.j	Other piles		No provision
	Preliminary and test piles	D31:1.*.*.1	No change
	Contiguous bored piles		Dealt with in Section D30

SMM6		SMM7	
Clause	Heading	Clause	Heading/Comment
	Raking piles	D31:1.*.*.2	Raking inclination ratio stated - No longer required to be stated in 10 degree increments (Refer also to Code of Procedure for Measurement for further explanation)
E.4.2	Above classifications to be grouped under sub-headings stating:-		No specific mention of grouping under sub-headings but this would still apply
	nominal cross section size	D31:1.1	Nominal diameter stated
	materials	D31:S1 & 2	Kind and quality of materials and tests
E.4.3	Piles other than cast-in-place concrete and interlocking steel piles		
E.4.3.a	The total number of piles	D31.1.1.1	Total number driven, specified length and commencing surface stated
E.4.3.b	The total driven length given in metres	D31.1.1.2	Total driven length
E.4.3.c	The total completed length		Specified length mentioned above - no longer given in ranges
E.4.4	Cast-in-place concrete piles		Dealt with in Section D30
E.4.5	Interlocking steel piles		Dealt with in Section D32

SMM6		SMM7	
Clause	Heading	Clause	Heading/Comment
E.4.6.a	Boring through rock		Not mentioned but if such work likely to occur whilst pre-boring then suggest measure in accordance with rules for D30:5.1
		D31:5 D31:M3	Redriving piles to be given as extra over piling and only where it is specifically required
E.4.6.b	Permanent casing		Not applicable
E.4.6.c	Placing concrete by tremie pipe		Not mentioned
E.4.7.a	Pre-boring where specifically required	D31:6 D31:C2	Maximum depth to be stated and Pre-boring is deemed to include grouting up voids between sides of piles and bores
E.4.7.b	Backfilling empty bores		Dealt with in Section D30
E.4.7.c	Jetting	D31:7	No change
E.4.7.d	Filling hollow piles	D31:8 & D31:S5	No change
E.4.7.e	Pile extensions	D31:9.1.1	Total number of extensions to be stated.
		D31:9.1.2 & 3	Extension lengths to be stated $\leq$ 3.00m $>$ 3.00m
		D31:C3	Preparing heads to receive pile extensions is deemed to be included
E.4.7.f	Cutting interlocking piles		Dealt with in Section D32

SMM6		SMM7	
Clause	Heading	Clause	Heading/Comment
E.4.8.e	Cuttting off tops of isolated piles	D31:10	Now to be given in metres stating number
		D31:C5	Now deemed to include for cutting off tops of piles and is deemed to include for preparation and integration of reinforcement into pile cap or ground beam and disposal. Code of Procedure for Measurement states that the responsibility as between main contractor and sub-contractor for cutting off tops of piles and for disposal of materials should be made clear in the Bills of Quantities. Cutting of tops of test piles should be given separately
E.4.8.b	Preparing pile heads	D31:C3	Preparing heads to receive pile extensions is deemed to be included
E.4.8.c	Forming enlarged bases		Not applicable
E.4.8.d	Heads and shoes	D31:C1	Driving heads and shoes are deemed to be included
E.4.8.e	Removal of piles		Not mentioned
E.4.8.f	Cutting steel interlocking piles to form holes		Dealt with in Section D32
E.5	Reinforcement	D31:1-3 D31:8.*.2	Details to be given in the description of the pile or work

SMM6		SMM7	
Clause	Heading	Clause	Heading/Comment
E.6	Disposal	D31:11	Now required to state whether disposed of on or off site. Specified locations and handling details to be stated.
		D31:M5	Method of calculating volume defined
E.7	Delays		
E.7.1	Standing time for rigs	D31:12.1 D31:M6 D31:C5	No change
E.7.2	Boring through artificial obstructions		Not mentioned but if such work likely to occur whilst pre-boring then suggest measure in accordance with D30:5.1
E.8	Tests	D31:13	Details of testing to be stated. No specific tests are now mentioned
E.9	Quantities		Generally dealt with under measurement and coverage rules as defined in the foregoing comments and as follows:-
E.9.1.a	Driven or bored length	D31:M1 & M2	Total driven depth includes for driving extended piles. Driving depth is measured along the axis of the pile from commencing surface to bottom
		D31:D1	of pile toe and is the depth specifically required by the designer
E.9.1.b	Concreted lengths		Not applicable

SMM6		SMM7	
Clause	Heading	Clause	Heading/Comment
E.9.1.c	Length for pre-formed piles	D31:D1	As above
E.9.1.d	Area for interlocking steel piles		Dealt with in Section D32
E.9.1.e	Pile extensions	D31:9	See above
E.9.1.f	Volume of surplus excavated material	D31:M4	Volume calculated from the nominal cross-sectional size of piles and their depth measures as D31:1-4.1.2.*

SMM6		SMM7	
Clause	Heading	Clause	Heading/Comment
	<u>Piling</u>	D32	<u>Steel piling</u>
E.1	Information	D31:P1 (a)-(c)	No change
		(d)	The relationship to adjacent buildings to be stated
E.2.1 & 2	Soil description	D32:P2 (a) & (b)	No change
E.2.3.		D32:P3(a)	No change
E.3	Plant	A43	Dealt with in preliminaries
E.4	Piling description		
E.4.1.a & b	Bored and driven cast-in-place concrete piles		Dealt with in Section D30
E.4.1.c	Pre-formed concrete piles		Dealt with in Section D31
E.4.1.d	Pre-formed prestressed concrete piles		Dealt with in Section D31
E.4.1.e	Pre-formed concrete sheet piles		Dealt with in Section D31
E.4.1.f	Timber piles		Not mentioned
E.4.1.g	Isolated steel piles	D32:1.1	Isolated piles
E.4.1.h	Interlocking steel piles	D32:2	Interlocking piles
E.4.1.j	Other piles		No provision
	Preliminary and test piles	D32:1.*.*.1	No change
	Contiguous bored piles		Dealt with in Section D30

SMM6		SMM7	
Clause	Heading	Clause	Heading/Comment
	Raking piles	D32:1.*.*.1.2	Raking inclination ratio stated - No longer required to be stated in 10 degree increments (Refer also to Code of Procedure for Measurement for further explanation)
E.4.2	Above classifications to be grouped under sub-headings stating:-		No specific mention of grouping under sub-headings but this would still apply
	nominal cross sectional size	D32:2.1	Section modulus and cross-sectional size, or section reference stated
	materials	D32:S1 & 2	Kind and quality of materials and tests
E.4.3	Piles other than cast-in-place concrete and interlocking steel piles		Dealt with in Section D31
E.4.4	Cast-in-place concrete piles		Dealt with in Section D30
E.4.5	Interlocking steel piles		
E.4.5. a & b	The total driven area	D32:2.1.1-3	Total areas to be specified in lengths $\leq$ 14.00m 14.00 - 24.00m < 24.00m
		D32:2.1.4	Total driven area
		D32:M4	Method of calculating areas of interlocking piles
E.4.5.c	Corners etc.	D32:3.1-4 D32:M5	No change The length measured for items extra over is the total length

SMM6		SMM7	
Clause	Heading	Clause	Heading/Comment
E.4.6. a-c	Extra over items		Dealt with in Section D31
E.4.7 a-d			Not applicable
E.4.7.e	Pile extensions	D32:4	Isolated pile extensions
		D32:5	Interlocking pile extensions
		D32:4-5. 1.1	Total number of extensions to be stated
		D32:4-5. 1-3	Extension lengths to be stated in $\leq$ 3.00m lengths $>$ 3.00m
		D32:4-5. *.*.1-4	To be measured stating whether to .1 Preliminary piles .2 Raking, inclination ratio .3 To be extracted .4 Using materials arising from cutting off surplus lengths of other piles
		D32:M6	Separate items to be given for the length of pile extensions and for the number of pile extensions
		D32:C2	Cost of extraction deemed to be included with piles so described
		D32:C3	Pile extensions are deemed to include the work necessary to attach the extension to the pile
			Note: the previous items in connection with isolated pile extensions equally apply to interlocking pile extensions

SMM6		SMM7	
Clause	Heading	Clause	Heading/Comment
E.4.7.f	Cutting interlocking piles to remove tops	D32:6.2.2	To be measured in linear metres stating the number
		D32:6.2.2.1	To be measured stating whether to preliminary or raking piles stating inclination ratio
		D32:C4	Deemed to include provision and filling of working space and disposal
E.4.8.a	Cutting off tops of isolated piles	D32:6.1.1.*	Refer to notes for cutting off tops of interlocking piles
E.4.8.b	Preparing concrete pile heads		Dealt with in Section D31
E.4.8.c	Forming enlarged bases		Not applicable
E.4.8.d	Heads and shoes		Dealt with in Section D31
E.4.8.e	Removal of piles	D32:C1	The cost of extraction is deemed to be included with piles so described
E.4.8.f	Cutting steel interlocking piles to form holes	D32:7.1	Enumerated with a dimensioned description
E.5	Reinforcement		Not applicable
E.6	Disposal		Not applicable
E.7	Delays		
E.7.1	Standing time for rigs	D32:8.1 D32:M8 D32:C5 D32.8.1.1 & 2	Delays to be described for isolated piles and interlocking piles and the time given in hours
E.7.2	Boring		Not applicable

SMM6		SMM7	
Clause	Heading	Clause	Heading/Comment
E.8	Tests	D32:13	Details of testing to be stated. No specific tests are now mentioned
E.9	Quantities		Generally dealt with under measurement and coverage rules as detailed in the foregoing comments and as follows:-
E.9.1.a	Driven or bored length	D32:M1 & M2	Total driven depth includes for driving extended piles. Driven depth is measured along the axis of the pile from commencing surface to bottom of pile toe
E.9.1.b	Concreted lengths		Not applicable
E.9.1.c	Length for pre-formed piles	D32:D1	The specified length is that specifically required by the designer
E.9.1.d	Area for interlocking steel piles	D32:M4	See above
E.9.1.e	Pile extensions	D32:4-5	See above
E.9.1.f	Volume of surplus excavated material		Not applicable

SMM6		SMM7	
Clause	Heading	Clause	Heading/Comment
E.10-13	**Diaphragm Walling**	D40	**Diaphragm Walling**
E.10	Information	D40:P1(a) & (b)	No change
E.11	Soil description	D40:P2	Substantially unchanged but different wording
E.12	Plant	A43	Dealt with in preliminaries
E.13.2, 3 and 5	Excavation and Disposal	D40:1.1 D40:S1 & 2	Maximum depth to be stated but excavation not given in stages as SMM6. Disposal now not given as a separate item. Details of support fluid and limitations of method of disposal to be given
E.13.4	Excavation in rock and artificial hard material	D40:2	No longer an alternative to be measured as extra over excavtion. No reference to standing time and associated labour being deemed to be included in respect of such items
E.13.6	Standing time and associated labour	D40:11	Delays - unchanged
E.13.7	Concrete in diaphragm walls	D40:M2(c)	No restriction in size of cast in accessories when calculating volume of concrete
E.13.8	Trimming and cleaning faces of diaphragm walls	D40:7	No change

SMM6		SMM7	
Clause	Heading	Clause	Heading/Comment
E.13.9	Reinforcement	D40:5	Unchanged apart from revisions in method of measurement of reinforcement noted under Section E30
E.13.10	Formwork		No reference to formwork. We consider this to be an error and suggest that where formwork is required it should be measured in accordance with Section E20
E.13.11	Waterproof joints	D40:8	No change
E.13.12	Guide walls	D40:9	To be stated whether to one side or both sides
E.13.13	Preparing tops of walls	D40:6	Cutting off to specified level
		D40:C1	provision and filling of working space is deemed to be included
	Protection		
E.14	Protecting the work		Dealt with in Preliminaries under "Employer's requirements" and "Contractor's general cost items"
		A34:1:6	
		A42:1.11	

62

SMM6		SMM7	
Clause	Heading	Clause	Heading/Comment
There are no comparable items in SMM6 for the following:-			
			Items of ancillary work in connection with diaphragm walling now to be given as follows:-
		D40:10.1	Preparing cast in pockets or chases at junctions
		D40:10.2	Excavating temporary backfill
		D40:10.3	Removal of guide walls

SMM6		SMM7	
Clause	Heading	Clause	Heading/Comment
F	CONCRETE WORK Generally	E	IN SITU CONCRETE/ LARGE PRECAST CONCRETE
		E10 E11 Q21	In situ concrete Gun applied concrete In situ concrete roads/pavings/bases Work in Q21 is measured in accordance with the Work Sections E10, E20, E30, E40, E41 and E42 as appropriate
F.1	Information	E10:P1	Information Reworded with more details in measurement rules, coverage rules and supplementary information
F.2	Plant	A43	Dealt with in preliminaries
F.3.1	Classification of work	-	No requirement to keep concrete framed structures, steel framed structures and other concrete work separate
F.3.2	Concrete volumes	-	No requirement for the volume of insitu concrete to be stated
	In-situ concrete		
F.4	Generally	E10:M1 E10:C1 E10:S1-5	All general notes are given at the beginning of the concrete section
F.5	Classification of size		

SMM6		SMM7	
Clause	Heading	Clause	Heading/Comment
F.5.1	Sectional areas	-	No classification of sectional areas is given or required
F.5.2	Thicknesses	E10:4-8	The classification of thickness for insitu concrete has been amended to:
		E10:4-8.1	$\leq$ 150mm
		E10:4-8.2	150 - 450mm
		E10:4-8.3	> 450mm
F.5.3	Thicknesses exclude projections or recesses	E10:M2	Unchanged
F.6	Concrete categories		
F.6.1	All in-situ concrete unless otherwise stated is measured in cubic metres	E10:1-13	All in situ concrete components ares measured in cubic metres
F.6.2	Foundations in trenches	E10:1 (D1)	No requirement to state "in trenches" or the thickness. Now includes atttached pile caps as well as attached column bases
F.6.3	Isolated foundation bases to columns and piers	E10:3 (D2)	Now includes machine bases and isolated pile caps. There is no requirement to state the number
F.6.4	Casings to steel grillages	E10:1 or E10:3	No separate item required. Assumed these will be included with E10.1 or E10.3 as appropriate
F.6.5	Ground beams	E10:2	No requirement to state the cross-sectional area

SMM6		SMM7	
Clause	Heading	Clause	Heading/Comment
F.6.6	Casings to steel ground beams	E10.2	These will be included with Ground beams
F.6.7	Pile caps	E10:1 or E10:3 (D1 & 2)	Now included with Foundations or Isolated Foundations as appropriate
F.6.8	Beds	E10:4 (D3)	No change in requirements but thickness ranges amended. Beds include blinding beds, plinths and thickenings of beds
F.6.9	Suspended slabs	E10:5 (D5)	Slabs - no longer required to say "suspended" and now also includes "column drop heads" - thickness ranges amended
F.6.10	Coffered or trough slabs	E10:6	No change except thickness ranges amended
F.6.11	Upstands and kerbs	E10:14	No requirement to state cross-sectional area
F.6.12	Walls	E10:7	No requirement to keep "walls not exceeding 1.50m high", "manhole walls" and "walls of horizontal or sloping ducts occurring in beds" separate from other walls. Retaining walls will also be measured and described as "Walls"

SMM6		SMM7	
Clause	Heading	Clause	Heading/Comment
F.6.13 and F6.14	Isolated beams and casings to isolated steel beams	E10:9.1 and E10:10.1	Beams, isolated Beam casings, isolated Now separate items but no requirement to state cross-sectional area or the number or the number
	Deep beams and deep casings to steel beams	E10:9.2 E10:9.3 and E10:10.2 E10:10.3	Beams, isolated deep Beams, attached deep Beam casings, isolated deep Beam casings, attached deep Now separate items but no requirement to state cross-sectional area
F.6.15	Isolated columns and isolated casings to steel columns	E10:11 and E10:12	Columns Column casings Now separate items but no requirement to state cross-sectional area or the number
F.6.16	Steps and Staircases	E10:13	Staircases No change
F.6.17	Tops and cheeks of dormers	- E10:5 and E10:7	No longer a separate item. Surveyor will have to decide whether to include with "slabs" and "walls" or whether to set up a special description dependant upon design and circumstances
F.6.18	Filling to hollow walls	E10:8	No change in requirements but thickness ranges amended
F.6.19	Machine bases	E10:3 (D2)	Now included with "isolated foundations" with no requirement to state the number

SMM6		SMM7	
Clause	Heading	Clause	Heading/Comment
F.6.20 and F.6.21	Filling to pockets	E10:17	All mortices are enumerated and all holes are measured in cubic metres stating the number, both irrespective of size
–	–	E11	**Gun applied concrete**
F.4.5	No separate provision for gun applied work but F.4.5 would have covered this	E11:1-4	To slabs, walls, beams and columns
		E11:M1	Reinforcement in reinforced gun applied concrete is measured with all other reinforcement under E30
		E30	
F.7	Joints	E40	**Designed joints in in situ concrete**
F.7.1	Day joints	E40:M1	No change
F.7.2	Designed joints	E40:1-3	Measured in 150mm stages
		E40:4	Sealants - now stated separately from the designed joint giving the size and the material
F.7.3	Welded or purpose made angles or intersections of water stops, etc.	E40:5 and E40:6	No change
		E41	**Worked finishes/ Cutting to in situ concrete**
F.8	Finishes cast on to concrete	E41:7	Finishings achieved by other means - measured in square metres stating type of finish, how achieved and position of work
F.9	Labours on concrete	E41	

68

SMM6		SMM7	
Clause	Heading	Clause	Heading/Comment
F.9.1	Labours	E41:1-11	No requirement to state whether on set or unset work - obvious from type of finish required
F.9.2	Curved labours	E41:M1	Curved work to be so described
F.9.3	Treating surface of unset concrete	E41:1-3	No change
F.9.4	Working concrete around pipes etc. of panel heating systems	E10:15	To be measured as extra over the work in which it occurs
F.9.5	Hacking faces	E41:4	No requirement to state purpose for which key is required
F.9.6	Grinding, sand-blasting, etc.	E41:5-6	No requirement to keep work over 3.50m high separate. No mention of work with margins of a different finish but it would seem sensible to make mention of this if it occurs
F.9.7	Formation of channels, chases etc.	E20	Formwork
F.9.8	Channels and chases required to be cut	E41:8-9	No requirement for a bill diagram. Cutting chases and rebates are to be measured stating
		E41:M2	the width only if a specific width is required. The depth is to be stated as $\leq$ 50mm, 50-100mm, 100-150mm and thereafter by giving

SMM6		SMM7	
Clause	Heading	Clause	Heading/Comment
		E41:8-9.* .*.3	the exact depth. Work in reinforced concrete shall be so described
F.9.9	Mortices, pockets and holes	E41:10-11	No requirement for a bill diagram. Cutting mortices and holes are to be measured stating the cross-sectional size only if a specific
		E41:M3	size is required. The depth is to be stated as ≤ 100mm, 100-200mm, 200-300mm and thereafter the exact depth is to be
		E41:10-11 *.*.3	given. Work in reinforced concrete shall be so described
F.9.10	Making good	E41:8-11.* .*.2	No change - include with item where required
F.10	Sundries		
F.10.1	Grouting under steel bases or grillages	E10:16	Grouting is now to be enumerated. No requirement to state size or thickness
F.10.2	Anchor bolts and other fixing devices	E42	**Accessories cast into in situ concrete** To be measured superficial, linear or enumerated as appropriate, giving all necessary information
	Reinforcement	E30	**Reinforcement for in situ concrete**
		E30:P1	The relative positions of all reinforcement shall be shown on drawings accompanying the Bills of Quantities

SMM6		SMM7	
Clause	Heading	Clause	Heading/Comment
F.11	Bar reinforcement	E30:1	Little change except there is no requirement to separate reinforcement into the various positions in which it is to be fixed. Only required to state the nominal size, whether in links, straight or bent bars or curved bars and whether horizontal or vertical. Measurement of horizontal bars over 12.00m long remains unchanged - they shall have lengths stated in stages of 3.00m. Measurement of long vertical bars is amended. The length above 6.00m is to be stated in stages of 3.00m
		E30:2	Spacers, chairs and special joints are measured <u>only</u> where
		E30:M3	they are not at the discretion of the Contractor.
F.12	Fabric reinforcement	E30:4	Fabric - Little change except raking and curved cutting is
F.12.6			not stated as being measurable. There is no requirement to
F.12.7			keep separate the bending of mesh whose bars exceed 4mm diameter No requirement to state position being fixed in strip reinforcement now superficial but width still stated

SMM6		SMM7	
Clause	Heading	Clause	Heading/Comment
F.12.5			Self-centering fabric is not mentioned but we consider should be measured separately from other fabric and temporary strutting would be stated
	Formwork		
F.13	Generally	E20	**Formwork for in situ concrete**
F.13.1	Categories	E20:1-28	Some change in categories, types and wording, see later detail
F.13.3	No deduction for voids not exceeding 5.00m2	E20:M4	This provision now only applies to formwork to:-
		E20:8	Soffits of slabs
		E20:9	Soffits of landings
		E20:10	Soffits of coffered or troughed slabs
		E20:11	Top formwork
		E20:12	Walls
F.13.4	Formwork coated with a retarding agent	-	This is not specifically mentioned but as it is an extra cost it will be necessary for this to be stated
F.13.5	Formwork left in shall be so described	E20:D2	No change in requirement but an explanation is given of the meaning of "Formwork left in" which is defined as that which is not designed to remain in position but is nonetheless impossible to remove

SMM6		SMM7	
Clause	Heading	Clause	Heading/Comment
		E20:D3	"Permanent formwork" is defined as that which is designed to remain in position
F.13.6	Formwork to curved surfaces	E20:M2	Radius to be stated - no mention of the geometric nature
F.13.7 and F.13.8	Formwork to soffits less than 1.00m high Formwork to soffits over 3.50m high	E20:8-10. *.*.1-2	Formwork to soffits to be given in height stages of "not exceeding 1.50m" and "thereafter in stages of 1.50m"
F.13.9	Requirements for special surface features	E20:20	Specially formed finishes are to be measured as "Extra over a basic finish and details of the specially formed finish are to be given
F.13.10	Raking and curved cutting	E20:C2	All cutting is deemed to be included
F.13.11	No distinction between different methods of forming recesses	-	Not specifically mentioned
F.13.12	Cutting and fitting around pipes, etc.	E20:C1	Adaptation to accommodate projecting pipes, reinforcing bars and the like is deemed to be included
F.14	Formwork to foundations and beds		
F.14.1	Formwork to edges and faces of foundations	E20:1	Sides of foundations - edges of beds are now to be included with sides of ground beams (no longer

SMM6		SMM7	
Clause	Heading	Clause	Heading/Comment
			deemed to be included with edges and faces of foundations)
F.14.2	Bill diagram	E20:P1	No longer specifically requested but all members and their sizes should be indicated on location and further drawings
F.14.3	Formwork to kickers of walls (See also F.15.6)	E20:21 E20:M15 E20:S4	No change in unit of measurement. The item is deemed to include both sides of the wall and if a specific height is required it shall be stated
F.15.1	Formwork to soffits of slabs, staircases and landings	E20:8 E20:9 E20:25	Formwork to soffits of staircases is now included with formwork to stairflights (E23).
		E20:M16	Formwork to stairflights is to be measured in metres between top and
		E20:C5	bottom nosings and is deemed to include soffits, risers and strings
		E20:8-9. *.*.1-2	Propping of formwork to soffits is to be stated in stages of 1.50m
		E20:D6	Formwork to soffits of landings occurring at floor levels is to be included with formwork to soffits of slabs.
	Nr of separate surfaces to be stated	E20:9	Number of separate surfaces only to be stated when measuring formwork to landings

SMM6		SMM7	
Clause	Heading	Clause	Heading/Comment
F.15.2	Formwork to soffits of solid slabs exceeding 200mm thick	E20:8-9. 1-2.*.*	No change in rules but no longer any mention of <u>solid</u> slabs. Propping height to be stated in 1.50m stages
F.15.3	Formwork to sloping upper surfaces of slabs over 15 degrees from horizontal	E20:11	Top formwork - No change in rules except that number of separate surfaces is no longer required to be stated
F.15.4	Formwork to soffits of coffered, troughed or similar shaped slabs	E20:10	No detailed change. Propping height to be stated in 1.50m stages

F.15.5	Formwork to:-
	  attached beams
	  attached beam casings
	  strings
	  upstand beams
	  recesses
	  projecting eaves
		E20:13
		E20:14
		
		E20:25
		E20:4
		E20:17
		-
		
		E20:25.C5
			Formwork to regular shaped attached beams and attached beam casings is to be measured in square metres. Formwork to irregular shaped attached beams and attached beam casings is to be measured in linear metres and referred to a dimensioned diagram
			
			A definition of regular shape is given in definition rule D10
			
			Propping heights in stages of 1.50m are to be stated for formwork to attached beams and attached beam casings
			
			Formwork to strings is no longer measured separately but is included in the item of formwork to stairflights

SMM6		SMM7	
Clause	Heading	Clause	Heading/Comment
		E20:M13	Recesses, nibs or rebates occurring in beam formwork are included in the measurement of that formwork
		E20:17	Recesses are measured as "Extra over" the formwork in which they occur and are measured in linear metres stating the number and giving a dimensioned description
			Formwork to projecting eaves is no longer specifically mentioned. Surveyors will have to measure in accordance with the rules most appropriate to the particular design.
F.15.6	Formwork to kickers of walls (See also F.14.3)	E20:21 E20:M15 E20:S4	No change in unit of measurement. The item is deemed to include both sides of the wall and if a specific height is required it shall stated
F.15.7	Formwork to complex or shaped members	E20:28	A dimensioned description is required and although not specifically requested a dimensioned diagram would probably be required
	Formwork to form mortices or holes	E20:26-27	The girth of a mortice or hole shall be stated stages:-

SMM6		SMM7	
Clause	Heading	Clause	Heading/Comment
			not exceeding 500mm 500mm - 1.00m thereafter in stages of 1.00m
			The depth of mortices and holes shall be stated in stages:-
			not exceeding 250mm 250 - 500mm 500mm - 1.00m actual depth if in excess of 1.00m
		E20:D12	Mortices include pockets
		E20:D13	A hole is defined as being less than or equal to 5.00 square metres
F.15.8	Formwork to edges of slabs, steps in tops of slabs, steps in soffits of slabs and risers of staircases	E20:3 E20:5 E20:6 E20:25	The height stages for the measurement of formwork to edges of slabs, steps in top surfaces and steps in soffits have been amended and now align with the measurement of formwork to sides of foundations; that is in stages
			not exceeding 250mm 250 - 500mm 500mm - 1.00m exceeding 1.00m
			Formwork exceeding 1.00m high to these items is measured in square metres
		E20:C5	Formwork to risers of staircases, edges of staircase flights and ends of risers abutting walls are not measured separately but are

SMM6		SMM7	
Clause	Heading	Clause	Heading/Comment
			deemed to be included in the measurement of formwork to stairflights
F.16	Formwork to walls and associated features		
F.16.1	Formwork to faces of walls	E20:12	No requirement to state the number of separate surfaces. Formwork to walls inside stair wells and lift wells no longer required to be stated separately
			Formwork to walls includes formwork to isolated columns and column casings whose length on plan is greater than four times the thickness
F.16.2	Vertical surfaces requiring formwork exceeding 3.50m high	E20:12.*.1 and E20:M9	The height stage has been changed to "exceeding 3.00m". The total wall surface of walls requiring formwork over 3.00m high shall be included in the item (i.e. from base to top and <u>not</u> from 3.00m and above)
F.16.3	Formwork to one side of a wall	E20:12.*.3	Wall thickness and background to other side to be stated
F.16.4	No deduction of formwork for kickers	E20:M10	No change

SMM6		SMM7	
Clause	Heading	Clause	Heading/Comment
F.16.5	Formwork to projections		Refer to the comments given for formwork to attached beams (SMM6 reference F.15.5 & SMM7 reference E20.13 etc.)
F.16.6	Formwork to pilasters (deemed to include attached columns and attached column casings) and recesses	E20:15-16. 2.*.*	Formwork to regular shaped columns and column casings is to be measured in square metres; that to irregular shapes in linear metres. A dimensioned diagram is required for irregular shaped columns and column casings
F.16.7	Formwork to complex or shaped members, isolated corbels, pilaster heads and to form mortices and holes	E20:28	Complex shapes - formwork shall be described with a dimensioned description and enumerated. This would include work to isolated corvels pilaster heads and the like

		E20:26 and E20:27	Mortices
			
			Holes
			
			Formwork to mortices and holes shall be enumerated, the description shall give the girth in and depth stages below
			
			Girth shall be stated as:-
			
			  not exceeding 500mm
			  500mm - 1.00m
			  thereafter in
			  stages of 1.00m

SMM6		SMM7	
Clause	Heading	Clause	Heading/Comment
			Depth shall be stated as:- not exceeding 250mm 250 - 500mm 500mm - 1.00m actual depth if over 1.00m
F.16.8	Formwork to ends, sloping tops or soffits of walls and also to perimeters of openings exceeding 4.00m girth	E20:23 E20.24	Formwork to wall ends, soffits and steps in walls, and to openings in walls is measured in linear metres up to 1.00m wide in width stages of:- not exceeding 250mm 250 - 500mm 500mm - 1.00m Over 1.00m wide the items above are measured in square metres
F.17	Formwork to isolated beams and columns		
F.17.1	Formwork to isolated beams and beam casings	E20:13.3 and E20:14.3	The comments given above for attached beams and beam casings apply (SMM6 reference F.15.5, and SMM7 references E20:13 and E20:14)
F.17.2	Formwork to isolated columns and column casings	E20:15.3 and E20:16.3 E20:D8	The comments given above for attached columns and column casings apply (SMM6 reference F.16.6 and SMM7 references E20:15 and E20:16.2) Isolated columns and column casings are defined as being less than four times their thickness in width. If their width is greater than

\	SMM6		SMM7	
Clause	Heading	Clause	Heading/Comment	
			four times their thickness then the formwork to them is measured as formwork to walls	
	Precast concrete			
F.18	Generally	E50	<u>Precast concrete large units</u>	
		F31	<u>Precast concrete sills/ lintels/copings/ features</u>	
		H40	<u>Glass reinforced cement cladding/features</u>	
		H50	<u>Precast concrete slab cladding/features</u>	
		K33	<u>Concrete/Terrazzo partitions</u>	
			Note:- References below to E50 apply equally to F31, H40, H50 and K33	
		E50:P1	Details of precast members showing stressing arrangements, full details of anchorages, ducts, sheathing and vents and the relative positions of members, their sizes, thicknesses and permissable loads shall be given on location drawings or other drawings accompanying the Bills of Quantities	
F.18.1	Precast units including floors shall be enumerated and given under an appropriate heading	E50:1	Precast work can be enumerated or measured in linear or square metres if the requirements of measurement rules E50:M1 and M2 are met. (i.e. if the length of items is at the Contractor's	

SMM6		SMM7	
Clause	Heading	Clause	Heading/Comment
			discretion or the items are of a standard length they can be measured in linear metres stating the number. In the case of floor units, if the members are of the same length they can be measured in square metres stating the length)
			An appropriate heading is not requested but this would depend upon the particular bill layout adopted
		E50:2	Where precast items are measured in linear or square metres all angles, fair ends, stoolings and other like items shall be measured as extra over the items on which they occur
F.18.2	Particulars to be given	E50:C1	This coverage rule seems somewhat muddled. It states that precast units are deemed to include moulds, reinforcement, bedding, fixings, temporary supports, cast-in accessories and pretensioning. Details of reinforcement have to be stated however under E50:1.*.*.1 as do cast-in accessories under E50:1.*.*.2. Beddings and fixings are required to be given as

SMM6		SMM7	
Clause	Heading	Clause	Heading/Comment
		E50:3	supplementary information (S4) Joints between units shall be measured in linear metres or enumerated giving details of the size and type of joint or sealant. Enumerated joints can be given in the description of the item to which they relate
F.18.3	Formwork or moulds deemed included	E50:C1	No change
F.18.4	Contractor designed precast work	–	Not mentioned in SMM7
F.18.5	Prestressed precast units measured in accordance with F.33-39	E50:S6	Details of any pretensioning shall be given in the description of precast work
F.19	Copings, kerbs, etc.	)Q10))	<u>Stone/Concrete/ Brick kerbs/ edging/channels</u>
F.20	Tiles or slabs	)Q10:1)))Q10:2-4)))))))))))))	Excavation is measured in accordance with Section D20 Kerbs/Edgings/ channels are to be measured in linear metres and the description shall include the size and extent of any reinforcement and details of the radius if curved. Foundation and haunching details are now also to be included in the description

SMM6		SMM7	
Clause	Heading	Clause	Heading/Comment
		Q10:3	Special units are measured as extra over the items on which they occur
		Q10.C1	All cutting is deemed included
		Q10:C2	Foundations/haunching is deemed to include formwork
			Refer to notes above (SMM6 reference F.18, SMM7 reference E50 etc.)
F.21	Standard or purpose-made units	H40	<u>Glass reinforced cement cladding/features</u>
		H50	<u>Precast concrete slab cladding/features</u> Both the above not specifically covered under SMM6 but generally as described under F.18 and F.21 Precast Concrete
		Q50	<u>Site/Street furniture/equipment</u> Precast concrete items of furniture/equipment are to be enumerated and fully described or referred to drawings
F.22	Temporary supports - details to be given	E50:C1	Temporary supports are deemed to be included
F.23	Cast stone to be in accordance with Section K	—	No mention in the precast section but would presumably be measured under F.22 - Cast stone walling/dressings

SMM6		SMM7	
Clause	Heading	Clause	Heading/Comment
F.24	Fixings	E50:S4	Fixings have to be given as supplementary information to items of precast work
	Composite Construction		
F.25	Generally	E60	**Precast/Composite concrete decking**
		E60:1	The in-situ and precast parts of composite slabs no longer have to be measured separately under a heading. One item of "Composite slabs" is required stating the overall thickness and is deemed to include solid concrete work and filling of ends. Margins up to 500mm wide are included with composite slabs whereas margins over 500mm wide are to be measured as in-situ concrete
	F.27 is very similar to these rules	E60:M1	
		E60:C1 E60:M2	
		E60:M3	
F.26 to F.32	**Hollow-block suspended construction**	-	Refer to composite construction above (SMM6 reference F.25, SMM7 reference E60)
F.33 to F.39	**Prestressed concrete work**		Refer to the appropriate SMM7 Work Section for the type of concrete
		E10 E50 E30 E20	i.e. In-situ concrete Precast concrete Reinforcement Formwork

SMM6		SMM7	
Clause	Heading	Clause	Heading/Comment
		E31	**Post tensioned reinforcement for in situ concrete**
F.33.7	Tendons tensioned after concrete cast	E31:1	Post tensioning is measured by the number of tendons in identical members
F.40 to F.44	**Contractor - designed construction**	-	Not specifically mentioned in SMM7 - measure in accordance with the appropriate rules and clearly state that it is Contractor - designed work
	Protection		
F.45	Protecting the work	A34:1.6 A42:1.11	Dealt with in Preliminaries under "Employer's requirements" and "Contractor's general cost items"

SMM6		SMM7	
Clause	Heading	Clause	Heading/Comment
G	**BRICKWORK & BLOCKWORK**	F	**MASONRY**
	Generally	F10	**Brick/Block walling**
		F11	**Glass block walling**
G.1	Information		Note: References below F10 apply equally to F11
G.1.1	General description of the work	A13	A general description is to be given in the preliminaries
G.1.2	Drawings to be provided	F10:.P1(a)	Plans and elevations to be provided together with additional requirements of principal sections showing position of and materials used in walls
		F10.P1(b)	External elevations are to be provided showing the materials used
G.2	Plant	A43	Dealt with in preliminaries
G.3	Classification of work into a) Foundations b) Loadbearing superstructure c) Non-loadbearing superstructures	–	The classification of work has been discontinued

SMM6		SMM7	
Clause	Heading	Clause	Heading/Comment
G.4	Measurement		
G.4.1	Brickwork and blockwork to be measured the mean length by the average height	F10:M1	Brickwork and blockwork to be measured on the centre line of the material unless otherwise required (ie. no change)
G.4.1.a	Voids not exceeding - 0.10m2 not deducted	F10.M2(a)	No change
G.4.1.b	Flues, lined flues and flue blocks - voids and work displaced not exceeding 0.25m2 not deducted	F10:M2(b)	No change
G.4.2	Labours on different kinds of work and on existing to be kept separate	F10:C1	Many labours are now deemed to be included:-
		F10:C1(b)	all rough and fair cutting
		F10:C1(c)	forming rough and fair grooves, throats. mortices, chases, rebates and holes, stops and mitres
		F10:C1(d)	raking out joints to form key
		F10:C1(e)	labours in eaves filling
		F10:C1(f)	labours in returns ends and angles
		F10:C1(g)	centering
G.4.3	Curved work to to be described stating radius	F10:M4	No change
G.4.4	Work in underpinning to be given as Section H	D50	Dealt with in Groundwork section

SMM6		SMM7	
Clause	Heading	Clause	Heading/Comment
	Brickwork		
G.5 G.5.1	Generally Particulars to be given	F10:S1-5	Additional requirements of S5 - Method of cutting where not at discretion of Contractor to be stated. S1 to S4 are as SMM6
G.5.2	Deduction of brickwork for string courses, lintels, plates and the like	F10:M3	No change in the requirements but 'sills' have been included in the examples
G.5.3	Classes of brickwork	-	The requirement that walls in trenches shall be given separately where the width of the trench does not exceed the thickness of the wall by more than 0.50m and deeper than 1.00m from the top of the excavation to the base of the wall has been discontinued
G.5.3a	Walls	F10:1.1	Now includes skins of hollow walls (D4). Walls are deemed vertical unless otherwise described (D3)
G.5.3b	Filling existing openings	C20:8	Dealt with in Alterations section and is to be given as an item with a dimensioned description and an identification
G.5.3c	Skins of hollow walls	F10:1 F10:D4	Now to be included with "Walls"

SMM6		SMM7	
Clause	Heading	Clause	Heading/Comment
G.5.3d	Dwarf support walls	F10:1.*.*.* F10:2.*.*.*	Not specifically mentioned - presumably will be measured as "Walls" or "Isolated piers"
G.5.3e	Projections of footings and chimney breasts	F10:5	Not specifically mentioned but plinths are measured as "projections" with attached piers etc - presumably projections of footings and chimney breasts will also be measured here
G.5.3f	Isolated piers and chimney stacks	F10:2 F10:4 and F10:D8	"Isolated piers" and "Chimney stacks" are to be measured separately. The definition of an isolated pier remains unchanged
G.5.3g	Battering walls	F10:1-4. *.2.* and F10:D5	The definition has been reworded but measurement remains unchanged
G.5.3h	Brickwork used as formwork. Temporary strutting given in description	F10:1-6.*. *.3	No change
G.5.3j	Backing to masonry. Cutting and bonding to masonry to be described	F10.1.*.*.2	Not specifically mentioned but could be measured as "Walls" with bonding to masonry described
G.5.3k	Refractory brick linings to flues	F10:9	Flue linings - No change. Non brick masonry flue linings are measured in Section F30:11.*.0.*

SMM6		SMM7	
Clause	Heading	Clause	Heading/Comment
G.5.3l	Brick damp-proof courses		Not mentioned separately - measure as "Walls" and add the additional information of "as damp-proof course" and number of courses
G.5.3m	Work in raising existing structures	F10:1	Not mentioned. Preparation of the top of the existing wall could be measured under Section C in linear metres and the new brickwork could be measured as "Walls"
G.5.4	Projections of attached piers, plinths, bands oversailing courses and the like	F10:5 F10:D9	No mention of "bands" but still to be measured in linear metres stating width and projection and now to state whether vertical, raking or horizontal
G.6	Thickening existing walls		
G.6.1a	Thickening existing walls	F10:1.*.*. 1 & 2	Not specifically mentioned. Measure as "Walls" stating building against and bonding to existing work"
G.6.1b	Projections on existing walls of attached piers, chimney breasts and the like	F10:5.*. *.1 & 2	Not specifcally mentioned. Measure as "Projections" built against and bonded to existing work
G.7	Tapering walls	F10:1.*. 3 - 4.*	No change

SMM6		SMM7	
Clause	Heading	Clause	Heading/Comment
G.8	Grooved bricks		Not mentioned but walls with all bricks grooved could be measured separately as "Walls" stating the type of brick (F10:S1). Where grooved bricks are to one side only of a wall (i.e. where part of the wall could be built in ordinary common bricks or bricks of another type) then the grooved bricks would be measured as "Extra over"
G.9	Cavities in hollow walls		
G.9.1	Forming cavities	F30:1	Still to be measured in square metres stating the width of the cavity and type, size and spacing of wall ties but now also to include rigid sheet cavity insulation where required
		F30:1.2.*.*	A new item of forming cavities between walls and other work has been included
G.9.2	Closing cavities	F10:12	No mention of ends, jambs or sills. Closing cavities are to be measured in linear metres stating the width of the cavity and method of closing together with whether the closing is vertical, raking, or horizontal

SMM6		SMM7	
Clause	Heading	Clause	Heading/Comment
G.10	Rough cutting	F10:C1(b)	All rough cutting is now deemed to be included
G.11	Rough chases	F10:C1(c)	All rough chases are now deemed to be included
G.12	Rough arches	F10:6	Not to be measured extra over as previously. To be measured the mean girth or length in linear metres stating height on face, width of exposed soffit and the shape of the arch
G.13	Bonding ends of walls	F10:25	No change in measurement. Extra material for bonding is now deemed to be included (C2)
	Brick facework		
G.14	Generally		
G.14.1	Rules for facework apply to fair face on brickwork	F10:D2	Facework is defined as any work in bricks or blocks finished fair
			Brickwork and Brick facework no longer have separate rules. The comments given above apply equally to brick facework. In addition the following comments apply specifically to facework
G.14.2	Particulars to be given	F10:S1-5	Additional requirements in S5 - Method of cutting where not at discretion of Contractor to be stated S1 to S4 are as SMM6

SMM6		SMM7	
Clause	Heading	Clause	Heading/Comment
G.14.3	Facework given as extra over	F10:1-5.2&3	Now to be measured full value stating whether to one or both sides
G.14.4	Deduction of facework	F10:M3	See above (SMM6 reference G.5.2)
G.14.5	Facework over half brick wide given in square metres	-	Not mentioned. Follow general measurement rules
G.14.6	Facework not exceeding half brick wide	F10:C1(f)	Labours in returns, ends and angles are deemed to be included.
		F10:11	Special bricks in reveals, intersections and angles are measured extra over the work which they occur
G.14.7	Facework built overhand to be described	F10:1-6.*.*.4	No change
G.14.8	Facework to panels and aprons not exceeding 1.00m2	-	Not mentioned. Measure according to the particular circumstances
G.14.9	Half brick walls and one brick walls built fair both sides or entirely of facings	F10:1-4.2&3	No change
G.14.10	Reveals and fair returns on walls entirely in facings or built fair both sides	F10:C1(f)	Returns are now deemed to be included.
		F10:11	Reveals, intersections and angles in special bricks are measured as extra over the work in which they occur
G.14.11	Battered facework	F10:1-4.*.2.*	Rate of batter no longer required to be stated

SMM6		SMM7	
Clause	Heading	Clause	Heading/Comment
G.14.12	Facework sunk or projecting less than half brick from general face	–	Not mentioned. Some reference would need to be made if this circumstance arose but could be measured to follow rule F10:3.2.*.* for ornamental bands but in square metres
G.14.13	Fair cutting and fair curved cutting	F10:C1(b)	All fair cutting is now deemed to be included
G.15	Fair angles and fair chases		
G.15.1	Fair vertical internal and external angles deemed to be included except to glazed brickwork	F10:C1(f) F10:26	Labours in returns, ends and angles are deemed to be included. Glazed brickwork is measured under "Surface treatments"
G.15.2	Fair battered internal and external angles	F10:C1 F10:11	All cutting and labours in forming angles are now deemed to be cinluded except where formed in special bricks
G.15.3	Fair squint and fair birdsmouth angles	F10:C1 F10:11	All as immediately above
G.15.4	Fair chamfered angles, fair rounded angles etc.	F10:C1 F10:11	All as immediately above
G.15.5	Fair chases	F10.C1(c)	All fair chases are now deemed to be included
G.16	Plain bands		

SMM6		SMM7	
Clause	Heading	Clause	Heading/Comment
G.16.1	Facework to flush plain bands not exceeding 300mm wide in facing bricks differing from the general facings	– F10:1	Not mentioned. Assume will be measured as "Walls" in a different brick
G.16.2	Facework to sunk or projecting plain bands	F10:5 F10:13	Measure projecting bands as projections stating width and depth of projection and whether vertical, raking or horizontal. Measure sunk bands in accordance with rules for "ornamental bands"
G.16.3	Fair ends and irregular angles	F10:C1	All fair ends and irregular angles are now deemed to be included
G.16.4	Plain bands over 300mm wide measured as walls	–	Not specifically mentioned. It would appear that all flush bands are measured as walls
G.17	Ornamental bands and cornices	F10:13	No change in measurement except that radii of curved work has to be stated. All ends and angles are now deemed to be included except where formed by special bricks
G.18	Tile creasing	F30:9	No longer measured as extra over the brickwork in which it occurs. To be measured as Accessories/Sundry items to brick/block/ stone walling. Ends, angles and pointing are now deemed to be included
G.19	Quoins	F10.14	No change

SMM6		SMM7	
Clause	Heading	Clause	Heading/Comment
G.20	Arches and tumblings		
G.20.1	Facework to arches	F10:6	No change in measurement. Centering is now deemed to be included
G.20.2	Facework to tumblings of buttresses	F10:19	No change
G.21	Sills, thresholds, copings and steps	F10:15-18	No change in measurement except ends and angles are deemed to be included. Angles formed with special bricks would be measured as extra over the work in which they occur
G.22	Key blocks, corbels, bases and cappings	F10:20-24	No change
G.23	Pavings	Q24 Q25	To be measured in accordance with Section Q - Pavings/Plantings/Fencing/Site furniture
	Brickwork in connection with boilers		
G.24	Boiler-seatings and boiler flues	F10:8 F10:9 F10:10	Boiler seatings Flue linings Boiler seating kerbs Boiler seatings and flue linings are to be measured in square metres stating the thickness. Curved work shall be stated with radius. Boiler seating kerbs are to be measured in linear metres

SMM6		SMM7	
Clause	Heading	Clause	Heading/Comment
			stating whether plain or irregular and the shape and size. Curved work shall be stated with the radius
G.25	Chimney Shafts	F10:7	Isolated chimney shafts are to be measured in square metres stating the number, size on plan, the shape, overall height, thickness and if built from outside scaffolding
	Blockwork		
G.26	Generally		Brickwork and blockwork no longer have separate rules. The comments given for brickwork above apply equally to blockwork. In addition the following comments apply specifically to blockwork
G.26.1	Particulars to be given		
G.26.2	Measurement of walls taken between attached piers. Piers are measured the combined thickness of the wall and pier	F10:1	Measurement of blockwork walls to be measured as for brickwork (i.e. overall - using mean vertical and horizontal dimenions) Attached piers are to be measured in in accordance with F10:5 stating the width and depth of projection

98

SMM6		SMM7	
Clause	Heading	Clause	Heading/Comment
G.26.6	Blockwork designed to be built without cutting blocks	-	Not specifically mentioned but would be required to be stated as supplementary information
G.27	Walls, partitions, piers and chimney stacks		
G.27.1	Walls and walls in trenches	-	No requirement to differentiate walls in trenches
G.27.1a	Walls and partitions	F10:1	The term used is now "Walls"
G.27.1b	Filling existing openings	C20:8	Dealt with in Alterations Section and is to be given as an item with a dimensioned description and an identification
G.27.1c	Skins of hollow walls	F10:1.D4	Now to be included with "Walls"
G.27.1d	Dwarf support walls	F10:1.*.*.* F10:2.*.*.*	Not specifically mentioned - presumably will be measured as "Walls" or "Isolated piers"
G.27.1e	Piers and chimney stacks	F10:2 and F10:4	Isolated piers and chimney stacks no longer to be grouped together
G.27.1f	Isolated casings	F10:3	Definition of an isolated casing no longer given - use discretion but suggest retain the SMM6 definition as a guide

SMM6		SMM7	
Clause	Heading	Clause	Heading/Comment
G.27.1g	Blockwork used as formwork. Temporary strutting given in description	F10:1-6. *.*.3	No change
G.27.1h	Work in raising existing structures	F10:1	Not mentioned. Preparation of the top of the existing wall could be measured under Alterations Section in linear metres and the new blockwork could be measured as "Walls"
G.27.2	Blockwork finished with a fair face and blockwork finished with facing blocks	F10:1-4.2&3	No change
G.27.3	Filling ends of hollow blocks or providing special blocks with solid ends - extra over	F10:11	Not specifically mentioned but could be measured "extra over" in accordance with rules for reveals etc.
G.27.4	Fair returns	F10:C1(f)	All fair returns are deemed to be included
G.28	Backing to masonry	F10:1.*.*.2	Not specifically mentioned but could be measured as "Walls" with bonding to masonry described.
G.29	Cavities in hollow walls	F30:1	Still to be measured in square metres stating the width of the cavity and type, size and spacing of wall ties but now also to include rigid sheet cavity insulation where required.

SMM6		SMM7	
Clause	Heading	Clause	Heading/Comment
		F30:1.2.*.*	A new items of "forming cavities between walls and other work" has been included.
G.30	Eaves filing		No change - no mention of blockwork designed to be built without cutting but would need to be stated as supplementary information
G.31	Rough cutting	F10:C1(b)	All rough cutting is now deemed to be included
G.32	Rough chases	F10:C1(c)	All rough chases are now deemed to be included
G.33	Bonding ends of walls	–	Not mentioned but as bonding superficially measured work to other work has to be stated we consider that bonding ends should also be measured and could be measured in accordance with F10:.25.*
G.34	Fair angles and fair chases	F10:C1	All fair angles and fair chases are now deemed to be included
G.35	Fair cutting		All fair cutting is now deemed to be included
G.36	Glass blockwork	F11	The general rules for blockwork apply equally to glass blockwork

SMM6		SMM7	
Clause	Heading	Clause	Heading/Comment
G.37	<u>Damp-proof courses</u>	F30:2	All damp-proof courses to be measured in square metres stating whether exceeding or not exceeding 225mm. No requirement to measure ends, angles and intersections on cavity gutters
G.38	Asphalt damp proof courses	J20:1	Now measured in Section J - Waterproofing
	<u>Sundries</u>		
G.39	Reinforcement	F30:3	No change
G.40	Keying - Raking out joints and hacking faces to form key	F10:Cl(d)	Now deemed to be included
G.41	Preparing for raising		Not mentioned - Assume will be measured in Alterations Section
G.42	Weather-fillets and angle-fillets	F30:4 & 5	No change
G.43	Bedding	F30:Cl(b)	Deemed to be included. Bedding and pointing windows/doors is measured in Section L - Windows/Doors/Stairs
G.44	Wedging and pinning up	F30:7	No change
G.45	Cutting grooves	F10:Cl	Cutting grooves is now deemed to be included
G.46	Expansion and designed construction joints	F30:8	No change

SMM6		SMM7	
Clause	Heading	Clause	Heading/Comment
G.47	Preparing for flashings and asphalt skirtings	F30:6	Pointing in flashings is measurable in linear metres. Cutting and forming grooves or chases are deemed to be included.
		F10:C1	Raking out joints is also deemed to be included
G.48	Building in or cutting and pinning	–	Not mentioned. Assume to be deemed included as with all rough and fair cutting around accessories (F30:C1a)
G.49	Holes	F10:C1	All holes are now deemed to be included
G.50	Mortices	F10:C1	Mortices are now deemed to be included
G.51	Making good	–	Deemed to be included
G.52	Air-bricks, ventilating gratings and soot doors	F30:12-14	To be enumerated separately. Each item is deemed to include forming of openings, liners, cavity closers and damp-proof courses
G.53	Flues	F30:11	Parging and coring flues no longer mentioned. No change in measurement of fireclay and precast concrete flue linings but cutting to form easings and bends and cutting to walls around linings is now deemed to be included
G.54	Gas flue blocks	F30:15	No change

SMM6		SMM7	
Clause	Heading	Clause	Heading/Comment
G.55	Chimney pots	F30:16	Not mentioned but could be included under proprietary items
G.56	Stoves and surrounds	F30:16	Comment as previous item
G.57	Centering	F10:C1(g)	Centering is deemed to be included
	Protection		
G.58	Protecting the works	A34:1.6 A42:1.11	Dealt with in preliminaries under "Employer's requirements" and "Contractor's general cost items"

SMM6		SMM7	
Clause	Heading	Clause	Heading/Comment
H	**UNDERPINNING**	D50	**UNDERPINNING**
	Generally		
H.1	Information		
H.1.2	Location and extent of work and particulars of existing structure	D50:P1(a) D50:P1(b)	Location and extent Details of existing structure
H.1.3	Information regarding nature of ground	D50:P2	To be in accordance with Section D20
H.1.4	Limit of length and number of sections	D50:P3	Unchanged
			Note:- Information to be provided is substantially the same
H.1.5	Work carried out from one or both sides	D50:*.*.*. 2-3	This is now a requirement of the rules for the various classifications of work in underpinning
H.1.6	Underpinning which is curved on plan	D50:*.*. *.1	As immediately above
H.2	Plant	A43	Dealt with in Preliminaries
	Work in all trades		
H.3	Excavation		
H.3.1	Temporary supports	D50:1	Unchanged except that details of making good are to be given
H.3.2	Working space allowances	D50:M1-3	Now referred to as width allowances which are substantially unchanged. See Measurement Rules M1

SMM6		SMM7	
Clause	Heading	Clause	Heading/Comment
			and 2. These width allowances would be added to preliminary trenches and underpinning pits when calculating the volume of these items
H.3.3	Excavation and filling of working space	D50:8	Filling now measured separately in accordance with D20:9 & 10.*.*.*
H.3.4	Excavation	D50:2.2	Excavating below the level of the base of the existing foundation now classified as "Underpinning Pits"
H.3.5	Cutting away projecting foundations	D50:5	Now to be classified as "Masonry" or "Concrete" with maximum width and depth of projection stated
H.3.6	Preparing the underside of existing work	D50:6	Unchanged
H.4	Disposal of water	D50:7 and D50:M7	Disposal. This clause now covers the disposal of excavated material as well as water and items are to be measured in accordance with D20:8.*.*.*
		D50:9	Surface treatments as Clause D20:13.*.*.*. Not previously mentioned but would have been included in accordance with rules for excavation

SMM6		SMM7	
Clause	Heading	Clause	Heading/Comment
H.5	Earthwork support	D50:4	Earthwork support is substantially unchanged and is measured in accordance with Clause D20:7.*.*.*
H.6	Concrete brickwork and asphalt	D50:10-14	Concrete, formwork, reinforcement, brickwork and tanking measured in accordance with appropriate Work Sections
	Protection		
H.7	Protecting the work		Dealt with in Preliminaries
		A34:1.6	under "Employer's requirements"
		A42:1.11	and "Contractor's general cost items"

SMM6		SMM7	
Clause	Heading	Clause	Heading/Comment
J	**RUBBLE WALLING** **Generally**	F20	**NATURAL STONE RUBBLE WALLING**
J.1	Information	F20:P1	No change in requirements for drawings and sequence of work but a general description is no longer requested
J.2	Plant	A43	Dealt with in Preliminaries
J.3	Measurement		
J.3.1	Templates and patterns deemed included	F20:C1(k)	No change
J.3.2	Measurement of superficial items	F20:M1 & 2	No change
J.3.3	Measurement of linear items	F20:M1 F20:M3	No change except that linear and enumerated items shall identify grooves, throats, flutes, rebates, cutting and mortices
J.3.4	Dry rubble work to be so described	F20:S1	No change
J.3.5	Curved work to be so described	F20:M4	No change
J.3.6	Pointing to be given in description	F20:S6	No change
J.3.7	Labours on existing rubble work to be so described	–	Not mentioned

SMM6		SMM7	
Clause	Heading	Clause	Heading/Comment
J.3.8	Dressed stonework	F20:C1(h)	Dressed margins to rubble work are deemed to be included
J.4	**Stone rubble work** Generally		As before except that walls are now to be measured total length and not between attached piers
		F20:4.D11	Piers are measured projection only
J.5	Walls, piers and chimney-stacks		
J.5.1a	Walls	F20:1	No change except walls in trenches no longer have to be kept separate
J.5.1b	Filling existing openings	C20:8	Now dealt with in Alterations section and is to be given as an item with a dimensioned description and identification
J.5.1c	Piers and chimney-stacks	F20:2-4	Now to be measured separately. Attached piers to have a dimensioned description stating the projection from the face of the wall. Isolated and attached columns are to be measured in linear metres with dimensioned descriptions
J.5.1d	Battering walls	F20:D5	No change
J.6	Tapering walls	F20:1-4. *.3 & 4.*	The rate of batter no longer has to be stated, otherwise no change

SMM6		SMM7	
Clause	Heading	Clause	Heading/Comment
J.7	Eaves filling	F20:C1(f)	No change
J.8	Bonding ends of walls	–	Not specifically mentioned but should be measurable
J.9	Rubble facework	F20:*.*.*. 4 & 5	All rubble walling which is faced is so described, either one side or both sides
J.10	Levelling uncoursed work	F20:C1(j)	Now deemed to be included
J.11	Cutting	F20:26 and F20:27	Rough and fair raking or circular cutting is measurable in linear metres.
		F20:C1(l)	Rough and fair square cutting is deemed to be included
J.12	Fair returns	F20:C1(g)	Labours in fair returns are deemed to be included
J.13	Fair angles	F20:C1(g)	Labour in fair angles are deemed to be included
J.14	Dressed margins	F20:C1(h)	Deemed to be included
J.15	Copings	F20:16	Horizontal, raking, vertical and curved copings are now measured separately
J.16	Arches and tumblings	F20:24	Arches can now be measured in linear metres or enumerated. If measured in linear metres then the number shall be stated. Arches are no longer to be measured as extra over the work in which they occur. Otherwise no change.
J.16.1	Arches		

SMM6		SMM7	
Clause	Heading	Clause	Heading/Comment
J.16.2	Tumblings	F20:33	Not specifically mentioned but would be measured as special purpose stones
	Sundries		
J.17	Grooves for water bars, flashings and skirtings	F20:28	Grooves are to be measured where on superficially measured work and included in the description of work measured in linear metres.
		F20:M3	
		F30:6	Pointing of flashings is measured separately
J.18	Chases	F20:32	Sizes of chases are now to be given in stages of girth - not exceeding 150mm - then in stages of 150mm
J.19	Building in or cutting and pinning	–	Not mentioned. Assume will be deemed to be included as with cutting for accessories (F30:C1)
J.20	Holes	F20:C1(c)	Deemed to be included
J.21	Mortices	F20:C1(c)	Deemed to be included except where linear items (M3)
J.22	Flues and chimney pots	–	As commented for SMM6 references G53, 54 and 55 (SMM7 references F30:11, F30:15 and F30:16)

SMM6		SMM7	
Clause	Heading	Clause	Heading/Comment

J.23	Centering	F20:36	To be measured. Stages for height of supports has changed to
			  3.00 - 4.50m
			    thereafter in stages of 1.50m
			No mention of centering to flat soffits but that to a flat arch etc. would be so described. All centering is to be enumerated

	Protection		
J.24	Protecting the work	A34:1.6 A42:1.11	Dealt with in preliminaries under "Employer's requirements" and "Contractor's general cost items"

SMM6		SMM7	
Clause	Heading	Clause	Heading/Comment
K	**MASONRY**	F21	**NATURAL STONE/ ASHLAR WALLING/ DRESSINGS**
		F22	**CAST STONE WALLING/ DRESSING** Note:- All references below to F21 apply equally to F22
K.1	Information	F21:P1	No change in requirements for drawings and sequence of work but a general description is no longer requested
K.2	Plant	A43	Dealt with in preliminaries
K3	Measurement		
K.3.1	Stone dressings	F21:D2	Definition of stone dressings
K.3.2	Templates and patterns deemed included	F21:C1(k)	No change
K.3.3	Superficial items - mean dimension		No change
K.3.4	Linear items - mean length		No change
K.3.5	Joggle joints etc. in description	F21:S7 & C1(a)	No change
K.3.6	Enumerated items	F21:33.M11	Special purpose stone - measurement rules unchanged
K.3.7	Plain, sunk circular and circular-circular work so described	F21:*.*.*. 10-13	No change
K.3.8	Stones or blocks over 1.50m long	F21:1-4.*. *.1	Only required to state that they exceed 1.50m long - no stages given in excess of 1.50m

SMM6		SMM7	
Clause	Heading	Clause	Heading/Comment
K.3.9	Stones over 0.50 cu.m	F21:1-4.*.*.2	Only required to state that a block exceeds 0.50 cu.m - no stages given in excess of 0.5 cu.m
K.3.10	Labours	F21:C1	Many labours are now deemed to be included
K.3.11	Pointing	F21:S6	No change
	Natural stonework		
K.4	Generally	F21:S1-S10	As SMM6 except that the following information is now also required.
		S2	Coatings to backs of stones.
		S10	Type and positioning of metal cramps, slates, dowels, metal dowels, lead plugs and the like
K.5.1a	Walls		Walls in trenches no longer have to be kept separate
K.5.1b	Filling existing openings	C20:8	Now dealt with in the Alterations section and is to be given as an item with a dimensioned description and identification
K.5.1c	Dwarf supports under seats, table-tops and the like	F21:33	Not now mentioned but if appropriate could be measured as special purpose stones and enumerated
K.5.1d	Piers and chimney-stacks	F21:2, 3 and 4	Columns and chimney-stacks now to be measured separately
K.5.1e	Battering walls	F21:1-4.*.2.* and D5	No change

SMM6		SMM7	
Clause	Heading	Clause	Heading/Comment
K.5.1f	Vaulting	F21:5	No change
K.6	Tapering walls	F21:1-4. *.3 & 4.* and D6	No change
K.7	Ends and angles of walls	F21:C1(g)	Labours in rough and fair returns, ends and angles are deemed to be included
K.8	Bonding ends of walls	-	Not specifically mentioned but should be measurable
K.9	Stone facework		
K.9.1	Stone facework bonded to backing of other material	F21:1-24. 1.*.8 and C3	No change
K.9.2	Stone facework built against backing of other material	F21:1-24. 1.*.6	No change
K.9.3	Stone facework used as formwork	F21:1-24. 1.*.15	No change
K.9.4	Stone facework built overhand	F21:1-24. *.*.18	No change
K.10	Stone slabbing or cladding to walls	H51	To be measured in accordance with Section H - Cladding/Covering
K.11	Rough cutting	F21:26 F21:C1(1)	Rough raking cutting and circular cutting is measurable. Other rough cutting is deemed to be included
K.12	Fair cutting	F21:27	Comments as for rough cutting above

SMM6		SMM7	
Clause	Heading	Clause	Heading/Comment
K.13	Rustications	F21:1-24. *.*.14	No longer a requirement to keep sunk and chamfered margins on granite, marble and slate separate otherwise no change
K.14	Grooves, rebates margins, flutes, sinkings and angles	F21:28-31 & M10	No change
K.15	Mouldings	F21:M7	Now included with band courses. Measurement rules unchanged
K.16	Enrichments	F21:M7	Now included with band courses. Measurement rules unchanged
K.17	Ornaments	F21:33	Not now mentioned by name but will be measured under "Special purpose stones" as stated in the Code of Procedure for Measurement
K.18	Small panels	-	As stated immediately above
K.19	Pilasters	F21:4; D10 & D11	Now referred to as "Attached columns" otherwise little changed
K.20	Quoins and jambs	F21:10 & 11	Measurement rules unchanged except attached, attached with different finish and isolated features have to be kept separate. Attached and isolated stones are defined in D12 and D13 respectively

SMM6		SMM7	
Clause	Heading	Clause	Heading/Comment
K.21	Columns	F21:3 & 4	Measurement rules unchanged but "Independent columns" now referred to as "Isolated columns". Caps and bases to columns are not mentioned by name but will be measured
		F21:33	as "Special purpose stones" (refer to Code of Procedure for Measurement)
K.22	Lintels, sills, mullions and transoms	F21:6-9	No change except ends are deemed to be included
K.23	Copings, cornices and band-courses	F21:14-16 D15	No change except ends are deemed to be included. Cornices are included with band courses
		F21:33	Kneeler blocks etc., are to be measured as "Special purpose stones"
K.24	Slab architraves and surrounds to openings	F21:12&13	No change. Plinth blocks, etc. are to be measured
		F21:33	as "Special purpose stones"
K.25	Arches	F21:24	Arches can be measured in linear metres or enumerated. If measured in linear metres the number must be stated. Springers, voussoirs and keystones to arches are to be
		F21:33	measured as "Special purpose stones"

SMM6		SMM7	
Clause	Heading	Clause	Heading/Comment
K.26	Tumblings to buttresses	)F21:33)	All to be measured as "Special purpose stones"
K.27	Pier caps and chimney-caps	))	
K.28	Finials, brackets and crobels	))	
K.29	Tracery	)	
K.30	Special features	)	
K.31	Steps, winders and landings	F21:21-23	No change except ends etc. are deemed to be included
K.32	Handrails, cappings and kerbs	F21:17-19 F21:33	No change. Angle blocks etc., will be measured as "Special purpose stones"
K.33	Balustrade panels	)F21:33)	To be measured as "Special purpose stones"
K.34	Baluster and newels	)	
K.35	Cover-stones and corbel-courses	F21:20 & 15 respectively	No change except ends are deemed to be included
K.36	Templates, bases and hearths	)F21:33)	All to be measured as "Special purpose stones"
K.37	Shelves, divisions, table tops and seats	)))	
K.38	Pavings	Q	To be measured in Section Q - Paving/Planting/Fencing/Site furniture

SMM6		SMM7	
Clause	Heading	Clause	Heading/Comment
K.39	Carving and sculpture	F21:34 & 35	Carvings and sculptures are to be enumerated stating the character of the work and giving a component detail drawing reference; the provision of any models is to be stated. Selecting blocks of size and quality, boasting for carving and working mouldings or similar members are all now deemed to be included
K.40 & 41	<u>Cast stonework</u>	F22	Cast stone walling/ dressings
			The rules given above for Natural Stone wall (F21) apply equally to Cast Stonework
K.42 & 43	<u>Clayware work</u>	F22	The rules for cast stonework apply
	<u>Sundries</u>		
K.44	Grooves for flashings and skirtings	F21:28	Grooves are only measured separately on superficial work. On linear and enumerated work grooves shall be identified in the description of the work.
		F30:6	Pointing grooves is to be measured as Accessories
K.45	Chases	F21:32	Chases are to be measured in linear metres stating the girth in stages of 150mm and whether rough or fair

SMM6		SMM7	
Clause	Heading	Clause	Heading/Comment
K.46	Holes	F21:C1	Holes are deemed to be included. Forming openings for gratings etc. are deemed to be included with the item
K.47	Mortices	F21:C1	Mortices (other than on linear and enumerated items) are deemed to be included. They shall be identified in the item on linear and enumerated items
K.48	Cramps	F21:S10	The type and positions of cramps etc. is to be given as supplementary information in the measurement of items
K.49	Coating backs of stones	F21:S2	To be given as supplementary information in the measurement of items
K.50 & 51	Centering	F21:36	To be enumerated stating whether to arches, tracery, projecting masonry or vaulting with a dimensioned description, also stating whether left in, to sloping soffits and/or the maximum height of supports of over 3.00m in stages of 1.50m. Definition rule D18 gives details of what the dimensioned description should include. Centering is deemed to include strutting, shoring, bolting, wedging, easing, striking, removing scribed and splayed edges as in

SMM6		SMM7	
Clause	Heading	Clause	Heading/Comment
			SMM6 and also all cutting and notching
	Protection		
K.52	Protecting the work		Dealt with in preliminaries under "Employer's requirements" and Contractor's general cost items"
		A34:1.6	
		A42:1.11	

SMM6		SMM7	
Clause	Heading	Clause	Heading/Comment
L	**ASPHALT WORK**	J	**WATERPROOFING**
		J20	**Mastic asphalt tanking/damp-proof membranes**
		J21	**Mastic asphalt roofing/insulation/ finishes**
		J22	**Proprietary roof decking with asphalt finish**
		J30	**Liquid applied tanking/damp-proof membranes**
		J31	**Liquid applied waterproof roof coatings**
		M11	**Mastic asphalt flooring**
			NOTE: Mastic asphalt roads are measured under Work Section Q22 - refer to in-situ finishings SMM6 clause T.4-12 herein
			NOTE: References below to J20 apply equally to J21, J22, J30, J31 and M11 unless stated otherwise
	Generally		
L.1	Information	J20:P1	Basic information requirement essentially unchanged from SMM6
L.2	Plant	A43	Dealt with in preliminaries
	Mastic asphalt		
L.3	Classification of work		
L.3.1a	Damp-proofing and tanking		Classification of work is as the above

SMM6		SMM7	
Clause	Heading	Clause	Heading/Comment
			work section headings measured as follows:
		J20:1	Tanking and damp-proofing
L.3.1b	Paving and sub-floors	M11:2	Flooring and underlay
L.3.1c	Roofing	J21:3 J22:3	Roofing
		J20:4	Paving
L.4	Measurement		Measurement rules except where unchanged noted below
		M11:M1	Mastic asphalt flooring in staircase areas and plant areas and plant rooms are each given separately. This was not an SMM6 requirement
L.4.2	No deduction for voids not exceeding 0.50m2	M3	No deduction is made for voids $\leq$ 1.00m2
L.4.3	Work not exceeding 300mm wide shall be given in metres stating the width in steps of 150mm	J20:1-4.4	All work is now to be measured in square metres and work is to be described with the following width ranges Width $\leq$ 150mm Width 150 - 225mm Width 225 - 300mm Width > 300mm
		J20:D1	Mastic asphalt flooring is deemed to be internal unless described as external
	Coverings are classified as flat, sloping over 10 but not exceeding 45 degrees from	J20:1-4.*.1	The pitch of the work is to be stated

123

SMM6		SMM7	
Clause	Heading	Clause	Heading/Comment
	horizontal and sloping over 45 degrees from horizontal and vertical		
L.4.4	Isolated areas not exceeding 1.00m2 shall be enumerated	-	Work in isolated areas is not mentioned
L.4.5	Underlay in contact with asphalt, and any reinforcement shall be given in the description	J20:S1	Kind, quality and size of materials including underlay and reinforcement
L.4.6	Work subsequently covered shall be so described	J20:1-4.*.*.1	No change
L.4.7	Work carried out overhand or in confined spaces shall be so described	J20:1-4.*.*.2	Overhand work - no change
		J20:1-4.*.*.3	Carried out in working space $\leq$ 600mm
L.4.8	Work and labours to retaining walls in sections	-	Not mentioned
L.4.9	Curved work, conical work, spherical work and elliptical work shall each be so described irrespective of radius	J20:M2	Curved work is so described. Conical, spherical or elliptical work is not mentioned but would need to be so described As the stating of curved work is a general measurement rule it would appear that curved labours will now have to be stated separately

124

SMM6		SMM7	
Clause	Heading	Clause	Heading/Comment
L.4.10	Internal angle fillets in linear metres and deemed to be two coat unless otherwise stated	J20:12	No change except dimensioned description must be given
L.5	Labours	J20:13 J20:14 J20:15 J20:16 J20:17	Fair edges Rounded edges Drips Arrises Turning asphalt nibs into grooves
L.5.3	Cutting to line and jointing new to existing asphalt shall be given separately in linear metres	J20:C1(a)	Work is deemed to include cutting to line
L.5.4 & L.5.5 and L.5.6	Working asphalt to metal or other flashings shall be given in linear metres. Working asphalt against frames of manhole-covers, duct-covers, mat-sinkings and the like shall be given in linear metres Working asphalt into outlet pipes, dishing to gullies shall be enumerated	J20:C2(a) & C1(c)	Working to metal or other flashings and working against frames of manhole covers, duct covers into outlet pipes and the like is now deemed to be included
L.6	Skirtings, fascias and aprons	J20:5 J20:6 J20:7	Skirtings Fascias Aprons

SMM6		SMM7	
Clause	Heading	Clause	Heading/Comment
L.6.1	Skirtings, fascias and aprons shall each be given separately in linear metres stating the width on face	J20:5-7. 1-3	No change except now to be measured in the following girth ranges: Girth ≤ 150mm Girth 150-225mm Girth 225 - 300mm Girth > 300mm girth stated
L.6.1	Fair edges, drips, arrises, internal angle fillets and turning nibs into grooves shall each be given in the description. Angles shall be enumerated	J20:4	Fair edges etc. are deemed to be included
L.6.2	Skirting on roof slopes shall be so described	-	Skirtings on roof slopes not mentioned - are presumably measured as "Skirtings; raking"
L.7	Gutters, channels, valleys and kerbs	J20:8 J20:9 J20:10 J20:11	Linings to gutters Linings to channels Linings to valleys Coverings to kerbs Rules as for skirtings, fascias and aprons apply
L.8	Cesspools and collars	J20:18 J20:19 J20:20	Collars around pipes, standards and like members Linings to cesspools Linings to sumps Rules essentially unchanged from SMM6 except arrises, internal angle fillets and outlets (where applicable) now deemed to be included Additional headings added:-

SMM6		SMM7	
Clause	Heading	Clause	Heading/Comment
		J20:21	Linings to manholes
		J20:22	Edge trim
L.9	Roof ventilators	J20:23	No change
	Asphalt tiling		
L.10	Asphalt tiling to be given in accordance with Section T		To be measured in accordance with J21 or M11 as appropriate
	Protection		
L.11	Protecting the work		Dealt with in preliminaries under "Employer's requirements" and "Contractor's general cost items"
		A34:1.6	
		A42:1.11	

SMM6		SMM7	
Clause	Heading	Clause	Heading/Comment
M	ROOFING	G	STRUCTURAL/CARCASSING METAL/TIMBER
		H	CLADDING/COVERING
		J	WATERPROOFING
		G30	Metal profiled sheet decking
		G31	Prefabricated timber unit decking
		G32	Edge supported/ Reinforced woodwool slab decking
		H30	Fibre cement profiled sheet cladding/ covering/siding
		H31	Metal profiled/flat sheet cladding/ covering/siding
		H32	Plastics profiled sheet cladding/ covering/siding
		H33	Bitumen and fibre profiled sheet cladding/covering/ siding
		H41	Glass reinforced plastics cladding/ features
		H60	Clay/concrete roof tiling
		H61	Fibre cement slating
		H62	Natural slating
		H63	Reconstructed stone slating/tiling
		H64	Timber shingling
		H70	Malleable metal sheet prebonded coverings/ cladding
		H71	Lead sheet coverings/ flashings
		H72	Aluminium sheet coverings/flashings
		H73	Copper sheet coverings/flashings
		H74	Zinc sheet coverings/ flashings

SMM6		SMM7	
Clause	Heading	Clause	Heading/Comment
		H75	<u>Stainless steel sheet coverings/flashings</u>
		H76	<u>Fibre bitumen thermoplastic sheet coverings/flashings</u>
		J40	<u>Flexible sheet tanking/damp-proof membranes</u>
		J41	<u>Built up felt roof coverings</u>
		J42	<u>Single layer plastics roof coverings</u>
		J43	<u>Proprietary roof decking with felt finish</u>
			The SMM6 Roofing Section has been divided into the above work sections and the appropriate SMM6 Roofing rules are related below to the relative SMM7 work sections
	<u>Generally</u>		
M.1	Information. Extent of work and height above ground to be indicated on location drawings	G30,31 and 32:P1(b)	Additional requirement for drawings to show the size of units to be stated where not at Contractor's discretion
		H30, 31, 32, 33, 41, 60, 61, 62, 63 and 64	No change
		H70, 71, 72, 73, 74, 75 and 76:P1(b)	Additional requirement for drawings to show the location of all laps, drips, welts, cross welts, beads, seams, rolls, upstands and downstands

SMM6		SMM7	
Clause	Heading	Clause	Heading/Comment
		J40, 41, 42 and 43:P1	Additional requirement for drawings to show any restrictions on the siting of plant and materials
M.2	Plant	A43	Dealt with in preliminaries
M.3	Classification of work		Refer to list of work sections given above
M.4	Measurement	H30, 31, 32, 33, 41, 60, 61, 62, 63, 64, 70, 71, 72, 73, 74, 75 and 76:M1 and J40, 41, 42 and 43:M2	No change except:- Voids $\leq$ 1.00m2 not deducted
			Apart from J40, 41, 42 and 43 no reference is made to measuring "the area in contact with the base" or "the area of the finished surface"
M.4.2	Work to dormers to be so described	H70-76:4	Work to dormers is only identified separately under H70 to H76
M.4.3	Work to curved, conical, elliptical and spherical roofs each so described		In all the above work sections curved work is so described stating the radii. Conical, elliptical and spherical work is not specifically mentioned but could be given as supplementary information

SMM6		SMM7	
Clause	Heading	Clause	Heading/Comment
M.5-17	<u>Slate or tile roofing</u>	H60	<u>Clay/concrete roof tiling</u>
		H61	<u>Fibre cement slating</u>
		H62	<u>Natural slating</u>
		H63	<u>Reconstructed stone slating/tiling</u>
		H64	<u>Timber shingling</u>
			Note: References below to H60: apply equally to H61, H62 and H64
M.5	Generally	H60:S1-4	No change
M.6	Roof coverings	H60:1-2	Measured superficially to roofs or walls stating pitch and if curved stating radii.
			Note: Voids $\leq$ 1.00m2 are not deducted and coverings are deemed to include battens and underlay, and work in forming voids $\leq$ 1.00m2 other than holes
M.7	Labours on roof coverings		
		H60:3	A linear item is now measured for abutments and is deemed to include all cutting
	Holes for pipes	H30:11	No change
	Forming small openings	C1b	Work forming voids $\leq$ 1.00m2 deemed to be included with coverings
M.8	Eaves	H60:4	Eaves
M.9	Verges	H60:5	Verges
			The above two items are classed as boundary work and measured in linear metres stating the

SMM6		SMM7	
Clause	Heading	Clause	Heading/Comment
		H60:C2	the method of forming, whether raking or curved work and are deemed to include undercloaks, bedding, pointing, ends, angles and intersections
			Note:- boundary work only measured to voids > 1.00m2
M.10	Valleys	H60:9	Valleys
M.11	Ridges, hips and vertical angles	H60:6 H60:7 H60:8	Ridges Hips Vertical angles
			The rules for ridges, hips, vertical angles and valleys do not state that they should be measured as "Extra over the roof coverings" but this would seem to be the most sensible approach. Cutting to ridges, hips, vertical angles and valleys is not given as a measurable item but neither is it stated as being deemed to be included. Coverage rule C2 for boundary work is not shown to apply to ridges, hips, vertical angles and valleys!
M.12	Special fillings		Not mentioned but would be included in the description of ridges, hips, vertical angles and valleys

SMM6		SMM7	
Clause	Heading	Clause	Heading/Comment
M.13	Hip irons	H60:10.4	Hip irons enumerated with a dimensioned description
M.14	Slates, soakers and saddles	H60:10.5-6	Soakers, saddles together with .1 Ventilators .2 Finials .3 Gas terminals are to be enumerated and fully described
M.15	Glass tiles		Not mentioned in SMM7
M.16	Lathing or battens	H60:Cl(a)	Deemed to be included with coverings
M.17	Underlay		
M.18-28	<u>Corrugated or troughed sheet roofing or cladding</u>	H30	<u>Fibre cement profiled sheet cladding/ covering/siding</u>
		H31	<u>Metal profiled/flat sheet cladding/ covering/siding</u>
		H32	<u>Plastics profiled sheet cladding/ covering/siding</u>
		H33	<u>Bitumen and fibre profiled sheet cladding/covering/ siding</u>
		H41	<u>Glass reinforced plastics cladding/ features</u>
			Note: References below to H30 apply equally to H31, H32, H33 and H41
M.18	Generally	H30:P1, S1 and S2	No change

SMM6		SMM7	
Clause	Heading	Clause	Heading/Comment
M.19.1 & 2	Roof coverings	H30:1	Measured superficially stating pitch and whether curved or fixed through underlining and are deemed to include integral underlays and work in forming voids $\leq 1.00m2$ apart from holes
M.19.3	Cranks and upstands	-	Not specifically mentioned but suggest could be measured extra over the coverings, similar to H30:18.6
M.20.1 & 2	Labours on roof coverings, cutting	H30:20	Cutting both raking or curved measured ion linear metres. Square cutting is deemed to be included
M.20.3	Holes for pipes etc.	H30:21	No change
M.20.4	Forming small openings	H30:C1	Voids $\leq 1.00m2$ are now deemed to be included
M.21	Filler-pieces	H30:17	Measured in linear metres stating dimensions and whether raking or curved (stating radii)
M.22	Bedding and pointing	H30:C2	Bedding and pointing on boundary work is deemed to be included together with ends, angles and intersections

SMM6		SMM7	
Clause	Heading	Clause	Heading/Comment
M.23	Ridges, hips and vertical angles	H30:6 H30:7 H30:8 H30:C2	Ridges Hips Vertical angles. On all ends, angles and intersections bedding and pointing are deemed to be included
M.24	Barge-boards	H30:11	Barge-boards - Angles and intersections as well as ends are now deemed to be included
M.25	Flashings and expansion joints	H30:10 H30:13	Expansion joints Flashings No change Note: boundary work is only measured to voids > 1.00m2
M.26	Louvres	H30:18.4 and H30:19.4	Sheets with louvre blades are enumerated stating the size as extra over the covering or cladding
M.27 and M.28	Roof-lights and special sheets Roof ventilators	H30:18 and H30:19	.1 Translucent sheets .2 Sheets with soaker flanges .3 Rooflight units .5 Ventilators .6 Junctions All are enumerated as extra over the roof or wall cladding with a dimensioned description
M.29-32	**Roof decking**	G30 G31 G32	**Metal profiled sheet decking** **Prefabricated timber unit decking** **Edge supported/ reinforced woodwool slab decking**

SMM6		SMM7	
Clause	Heading	Clause	Heading/Comment
		J42	<u>Single layer plastics roof coverings</u>
		J43	<u>Proprietary roof decking with felt finish</u>
			Note: References below to G30 apply equally to G31 and G32 unless otherwise stated
			References below to J42 apply equally to J43
			Work sections G31 and G32 follow similar rules to G30 except G31 also requires the following timber details:-
			a) whether sawn or wrot
			b) Selection and protection for subsequent treatment
			c) surface treatments
			d) matching grain or colour
			e) planing margins
M.29	Generally	G30:P1	Information required as SMM6 but must also state size of units where not at contractors discretion and surface treatments applied as part of the production process must also be stated
		G30:S1-3 J42	No change

SMM6		SMM7	
Clause	Heading	Clause	Heading/Comment
M.30	Roof coverings	G30:1 and G30:2	Measured in square metres and if measuring decking units the number of units is to be stated if not at the Contractors discretion
			Note: No deduction is made for voids not exceeding 0.50m2. No requirement to differentiate coverings over 50 degrees from horizontal and vertical
		J42:2.1	Plastics roof coverings and Proprietary roof decking with felt finish is measured in square metres, pitch stated
		J42:M2	Voids $\leq$ 1.00m2 are not deducted from the area measured for coverings
M.31	Labours on roof coverings	G30:3	Cutting is not mentioned and no guidance is given as to whether it should be measured or deemed to be included. We assume that it should be deemed to be included and an appropriate note should be made to that effect

SMM6		SMM7	
Clause	Heading	Clause	Heading/Comment
			Holes, notches, etc. are enumerated stating if on or off site and as extra over the decking or decking units
M.32	Bearings, eaves, kerbs and flashings	G30:4-10	Bearings, eaves, kerbs, abutments, nibs, blocks, fillets and profile fillers are all measured separately in linear metres
		G32:4-5	Woodwool kerbs, woodwool angle fillets are measured separately in linear metres
		G32:6	Filling rebates with insulating strips
		G32:7	Isolating strips
		G32:M3	Note: G32 to 7 are only measured under G32
M.31 and M.32	Labours on roof coverings Bearings, eaves, kerbs and flashings	J42:C2	Generally included in items below which are required to be measured as boundary work and are deemed to include all cutting, ends, angles, intersections, notching, bending, turning into grooves, wedging, dressing, trimming and jointing covering to flashings, working into channels and the like and filler pieces
		J42:3	Abutments
		J42:4	Eaves
		J42:5	Verges
		J42:6	Ridges
		J42:7	Hips
		J42:8	Vertical angles
		J42:9	Valleys
		J42:10	Skirtings
		J42:11	Flashings

SMM6		SMM7	
Clause	Heading	Clause	Heading/Comment
		J42:12	Aprons
		J42:13	Gutters and linings
		J42:14	Coverings to kerbs
		J42:3-14	The above items are measured in square metres where their girth exceeds 2.00m and in linear metres in stages of 200mm where their girth does not exceed 2.00m
M.31.4	Forming small openings (i.e. not exceeding 0.50m2) to be enumerated irrespective of size	J42:M3	Boundary work to voids is only measured where the void is > 1.00m2
M.33-39	<u>Bitumen-felt roofing</u>	J40	<u>Flexible sheet tanking/damp-proof membrane</u>
M.33	Generally		
		J41	<u>Built up felt roof coverings</u>
			Note: References below to J40 apply equally to J41
M.33.2	Underlay in contact with felt shall be given in the description	J40:S1	Kind, quality and size of materials including underlays required under supplementary information
		J40:1	Tanking and damp proofing
M.34	Roof coverings	J40:2	Roof coverings. All coverings now measured in square metres with the pitch stated
M.35	Labours	J40:3-14	Generally included in items below which are required to be measured as boundary

SMM6		SMM7	
Clause	Heading	Clause	Heading/Comment
			work and are deemed to include all cutting, ends, angles, intersections, notching, bending, turning into grooves, wedging, dressing, trimming and jointing covering to flashings, working into channels and the like and filler pieces.
M.35	Labours	J40:3 J40:4 J40:5 J40:6 J40:7 J40:8 J40:9	Abutments Eaves Verges Ridges Hips Vertical angles Valleys
		J40:3-14	The above items are measured in square metres where their girth exceeds 2.00m and in linear metres in stages of 200mm where their girth does not exceed 2.00m
M.36	Flashings, skirtings and aprons	J40:10 J40:11 J40:12	Skirtings Flashings Aprons
M.37	Gutters, valleys and kerbs	J40:13 J40:14	Gutters and linings Coverings to kerbs
M.38	Channels to be given in metres as extra over the coverings in which they occur	J40:C2	Working into channels is deemed to be included

SMM6		SMM7	
Clause	Heading	Clause	Heading/Comment
M.39	Cesspools and collars	J40:15 J40:16 J40:17	Linings to cesspools Linings to sumps Collars around pipes and standards No change The following are also to be measured under these main headings
		J40:18	Outlets and dishing to gullies (nr)
		J40:19	Edge trim (m)
		J40:20	Roof ventilators (nr)
		J40:21	Holes (nr)
		J40:22	Fire stops (m)
M.40-47 and M.48-55	<u>Sheet metal roofing</u> <u>Sheet metal flashings and gutters</u>	H70 H71 H72 H73 H74 H75 H76	<u>Malleable metal sheet prebonded coverings/ cladding</u> <u>Lead sheet coverings/ flashings</u> <u>Aluminium sheet coverings/flashings</u> <u>Copper sheet coverings/flashings</u> <u>Zinc sheet coverings/ flashings</u> <u>Stainless steel sheet coverings/flashings</u> <u>Fibre bitumen thermoplastic sheet coverings/flashings</u> Note: References below to H70 apply equally to H71, H72, H73, H74, H75 and H76
M.40	Generally	H70:S1-6	In addition to items required under SMM6 now need to state type of support materials and any special finishes

SMM6		SMM7	
Clause	Heading	Clause	Heading/Comment
		H70:P1	Also need to state the extent of roofing work and its height above ground and location of all laps, drips, welts, cross welts, beads, seams, rolls, upstands and downstands
M.40.3 a-g	Allowances made in calculating areas where none stated	H70:M2	Allowances are generally similar to those in SMM6 for lead but for all sheet coverings
M.41	Roof coverings	H70:1-9	Work now classified as roof coverings, wall coverings, preformed cladding panels, dormers, hoods, domes, spires, finials or soffits measured in square metres stating the pitch and if curved, the radii
M.42	Labours	H70:C1	Coverings now deemed to include: (a) isolated areas (b) work to falls and crossfalls (c) underlay in contact with the coverings (d) work in forming voids $\leq$ 1.00m2 other than holes Note: No deduction is made for voids $\leq$ 1.00m2 (M1) (e) dressing/wedging into grooves, hollows, recesses and the like

SMM6		SMM7	
Clause	Heading	Clause	Heading/Comment
M.42.1 M.42.2	Raking and curved cutting Notching, bending	)))	No longer required to be measured under SMM7
M.42.3	Welted and beaded edges etc.	H70:23	Only welted, beaded or shaped edges are to be measured
M.42.4	Dressing into hollows etc..	H70:C1(e)	Dressing/wedging into grooves, hollows, recesses and the like now deemed to be included with the coverings
M.42.5	Dressing over glass etc.	H70:24	Dressings to corrugated roofing, slating or tiling and glass and glazing bars are measured in linear metres stating the nature of the roofing and whether down or across corrugations
M.42.6 and M.42.7	Ends and dressing into outlet pipes	H70:C1	No longer required to be measured under SMM7 as deemed to be included with covering or flashing etc.
M.43	Nailing	H70:S3	Method of fixing should be stated in the description
M.44	Collars around pipes and standards	H70:29	Are enumerated stating size of member and length of collar and are deemed to include pipe sleeves
		H70:30	Holes for pipes, standards and the like are to be enumerated

SMM6		SMM7	
Clause	Heading	Clause	Heading/Comment
M.45	Cesspools and sumps	H70:20 H70:21 H70:22	Catchpits Sumps Outlets No change
M.46	Dots, ornaments and plugs	H70:31 H70:32 H70:33	Dots Ornaments Plugs No change
M.47	Underlay	H70:C1(c) H70:S1	Underlay in contact with covering is deemed to be included with the covering Other underlay is included in the description of the work
M.48	Generally	 H70:C2	See notes against SMM6.M.40 Generally Work is now deemed to include:- (a) laps, seams, ends (b) angles and intersections (c) rolls (d) upstands and downstands (e) dressing/wedging into grooves, hollows, recesses and the like
M.49 M.50 M.51	Profiles Flashings Preformed flashings	H70:10 H70:11 H70:12 H70:13 H70:14 H70:15 H70:16 H70:17 H70:18	Flashings Aprons Cills Weatherings Cappings Hips Kerbs Ridges Reveals, returns and jambs

SMM6		SMM7	
Clause	Heading	Clause	Heading/Comment
		H70:10-18. *.1-8	All are measured in linear metres with a dimensioned drawing or dimensioned description stating whether horizontal, sloping, vertical, stepped, preformed, dressed over corrugated roofing, slating, tiling, glass or glazing bars
M.50.2	Edges etc.	H70:23	Work to edges shall be measured in linear metres stating whether welted, beaded or shaped. Work to boundaries is only measured to voids when the void is > 1.00m2
M.52 M.53	Flat gutters Sloping gutters	H70:19	Gutters shall be measured in linear metres stating whether stepped, secret, sloping, tapered or preformed with either a dimensioned description or a dimensioned diagram
M.54 M.55	Soakers and metal slates Saddles	H70:25 H70:26 H70:27 H70:28	Saddles Soakers and slates Hatch covers Ventilators All are enumerated with a dimensioned description stating if handed to others for fixing
		H70:C3	Dressing and bossing is deemed to be included

SMM6		SMM7	
Clause	Heading	Clause	Heading/Comment
	Protection		
M.56	Protecting the work		Dealt with in Preliminaries under
		A34:1.6	"Employer's requirements" and
		A42:1.11	"Contractor's general cost items"

There is no comparable item in SMM6 for the following:-

		H30:22	Fire stops
			To be measured in linear metres with a dimensioned description

146

SMM6		SMM7	
Clause	Heading	Clause	Heading/Comment
N	<u>WOODWORK</u>		<u>Work Sections from the following Work Groups compare:</u>
		G	<u>STRUCTURAL/CARCASSING METAL/TIMBER</u>
		H	<u>CLADDING/COVERING</u>
		K	<u>LININGS/SHEATHING/ DRY PARTITIONING</u>
		L	<u>WINDOWS/DOORS/ STAIRS</u>
		N	<u>FURNITURE/EQUIPMENT</u>
		P	<u>BUILDING FABRIC SUNDRIES</u>
		Q	<u>PAVING/PLANTING/ FENCING/SITE FURNITURE</u>
	The SMM6 Woodwork trade has been fragmented into numerous Work Sections within SMM7. An easy comparison of the two documents is not possible. Whilst trying to keep to the SMM6 trade order we have had to deal in some instances with more than one set of SMM7 Work Sections under the same SMM6 trade sub-section. The same SMM6 references may therefore appear in more than one place below. SMM7 Work Sections applicable to one particular SMM6 sub-section follow each other in alphabetical sequence. SMM6 clause N.1 is applicable to the whole of the Woodwork trade. Where it has been necessary to refer back to this clause the clause numbers have been bracketed		
N.2-3	<u>Carcassing</u>	G20	<u>Carpentry/Timber framing/First fixing</u>
N.4-12	<u>First fixings</u>		
(N.1)	Information	G20:P1	Information to be provided has been reworded with more details in measurement, definition, coverage rules and supplementary information including the following variations from SMM6

147

SMM6		SMM7	
Clause	Heading	Clause	Heading/Comment
		G20:C1	1) Labours included in all members regardless of cross sectional area
		G20:S3	2) If fixing through vulnerable materials this should be noted
(N.1.4)	Items deemed to be fixed with nails. Other methods of fixing to be described	G20:S2	Method of fixing to be given where not at Contractor's discretion
(N.1.6)	Items in one continuous length; softwood exceeding 4.20m and hardwood exceeding 3.00m	G20:6-9.*.*.1	Length to be stated where required in one continuous length exceeding 6.00m
N.2	Carcassing items	G20:1-9	Classification of items has been enlarged
N.2.1	Classification of items		Trusses, trussed rafters, trussed beams, wall or partition panels and portal frames are all enumerated with dimensioned descriptions.
		G20:C2	Webs, gussets etc. are deemed to be included.
		G20:8	Plates are measured in linear metres and are to structural elements only and bearers.
N.2.2	Strutting and bridging between joists	G20:10	No change

SMM6		SMM7	
Clause	Heading	Clause	Heading/Comment
N.2.3	Cleats, sprockets and the like	G20:11-13	Three new items of butt jointed supports, framed supports and individual supports. Supports are defined as grounds, battens, firrings, fillets, drips, rolls, upstands, kerbs and the like
		G20:17-18 G20:D11	cleats; no change except they are defined as including sprockets and the like
N.3	Carcassing labours		
N.3.1	Labours	G20:C1	Labours now generally deemed to be included
N.3.2	Wrought surfaces	G20:19	Wrot surfaces are those only in sawn items with no change to the method of measurement
N.3.3	Cutting	G20:C1	Labours deemed to be included
N.3.4	Notching etc.	G20:18	Ornamental ends are enumerated
	First fixings (Note:- First Fixings - See also SMM7 Work Sections H20, H21, K11, K12, K20 and K21 following)		
N.4	Boardings and flooring		
N.4.1.a	Floors		Dealt with in Sections K11, K20 and K21
N.4.1.b	Walls		Dealt with in Sections H20 and H21

SMM6		SMM7	
Clause	Heading	Clause	Heading/Comment
N.4.1.c	Ceilings and beams		Dealt with in Section K20
N.4.1.d	Roofs		Dealt with in Section K20
N.4.1.e	Tops and cheeks of dormers		Dealt with in Section K20
N.4.1.f	Gutter boarding and sides	G20:14	Gutter boards - deemed to include sides. No change in measurement except widths over 300mm can be measured in square metres or in linear metres depending on cross-section shape
N.4.1.g	Eaves and verge boarding, fascias and barge boards	G20:15 G20:D10 G20:16	Fascia boards - deemed to include barge boards Eaves or verge soffit boards Measurement as for gutter boards
N.4.2-6 and N.5		-	Not applicable to G20
N.6	Firrings, drips and bearers	G20:11-13 G20:D7	To be measured as supports Supports are defined as including grounds, battens, firrings, fillets, drips, rolls, upstands and kerbs or the like
N.7-8		-	Not applicable to G20
N.9 to N.12	Fillets, rolls, grounds, battens and framework generally	G20:11-13 G20:D7	Supports Defined above

SMM6		SMM7	
Clause	Heading	Clause	Heading/Comment
	First fixings		
N.4-12	Boardings and flooring	H20	Rigid sheet cladding
		H21	Timber weatherboarding
		K11 $	Rigid sheet flooring/ sheathing/linings/ casings
		K12 $	Under purlin/inside rail panel linings
		K20 $	Timber board flooring/sheathing/ linings/casings
		K21 $	Timber narrow strip flooring/linings
			Note: References below to H20 below apply equally to H21, K11, K12, K20 and K21
			$ See also under SMM6N.15 - Sheet linings and casings where the same SMM7 Work Sections are applicable to "second fixings"
(N.1)	Generally	H20:D1	As SMM6 except:
			1. Work deemed to be internal unless described as external
		H20:C1(a)	2. All labours deemed to be included regardless of cross-sectional area
		H20:C1(b)	3. Breather paper lining/sheathing deemed to be included but given in description
		H20:S4-6	4. Details of preservative or fire retardant treatments should be

SMM6		SMM7	
Clause	Heading	Clause	Heading/Comment
		H20:S7	be described 5. Cover and jointing strips and cover moulding should be described
		H20:S11	6. Fixing through vulnerable finishes should be described
		H20:S13	7. Details of finish, trim or support should be described
N.4.1 a-e	Floors, walls, ceilings and beams, roofs, tops and cheeks of dormers	H20:1-7	No change except: 1. Beams and columns measured superficially in stages of 600mm girth starting at 600mm girth to the external finished girth Note: sloping defined as > 10 degrees from both horizontal or vertical
N.4.1.f	Gutter boarding and sides		Dealt with in Section G20
N.4.1.g	Eaves and verge boarding		Dealt with in Section G20
N.4.2	No addition for joints or laps and no deduction for voids not exceeding 0.50 square metres	H20:M1	Addition for joints and laps not mentioned. No change in deduction for voids
N.4.3	Members laid diagonally	H20:1-7. *.*.1	No change
N.4.4	Members not exceeding 1.00 metres in length	-	No longer required to be stated

SMM6		SMM7	
Clause	Heading	Clause	Heading/Comment
N.4.5	Areas not exceeding 300mm in width	H20:1-5.2	No change
N.4.6	Areas not exceeding 1.00 square metres	H20:1-5.3	No change
N.5	Labours on boarding and flooring	H20:C1	Deemed to be included except:
		H20:8 H20:D8	(1) Abutments (defined as being where detail is different from the standard detail)
		H20:9 H20:D9	(2) Finished angles, both internal or external (defined as where the decorative veneer etc. is returned or where on panelling angles are other than butt jointed)
		H20:10	(3) Holes for pipes, standards or the like
N.8	Access traps, cesspools, etc.	H20:12	No change
		H20:11	Fire stops are measured in linear metres

SMM6		SMM7	
Clause	Heading	Clause	Heading/Comment
	Second fixings	P20	Unframed isolated trims/skirtings/sundry items
N.13	Unframed second fixings	P20:M2	Items are only measured independently in this Section where not specified as part of another Work Section
(N.1)	As applicable to second fixings		
(N.1.1)	Particulars to be given	P20:S1-9	No change except additional item of "fixing through vulnerable materials" to be given as applicable
(N.1.2)	Sizes nominal unless stated as finished	P20:D1	No change
(N.1.3)	Method of jointing	P20:S7	No change
(N.1.4)	Items deemed fixed with nails unless stated	P20:S8	Method of fixing to be stated where not at the Contractor's discretion
(N.1.5)	Timber over 200mm in one width	-	Not mentioned
(N.1.6)	Items in one continuous length	-	Not mentioned
(N.1.7)	Labours to be related to the particular item	-	No change
(N.1.8)	Curved work so described	P20:M3	No change
(N.1.9)	Measurement of labours	-	No change
(N.1.10)	Ends, angles, mitres etc. deemed to be included except when	P20:C1	All ends, angles, mitres, intersections and the like are deemed to be included

SMM6		SMM7	
Clause	Heading	Clause	Heading/Comment
	cross-sectional area exceeds 0.002m2		except on <u>hardwood</u> items with a cross-sectional area exceeding 0.003m2
(N.1.11)	Number of different cross-sectional shapes to be given	P20:1-8. *.*.3	No change
		P20:M1	Items which do not have a constant cross-section are so described and given stating extreme dimensions
(N.1.12)	Girth of a mould over 100mm stated in 25mm stages	-	Not mentioned
N.13.1.a	Skirtings, picture rails, dado rails and the like	P20:1	Now includes architraves. Architraves as part of a door or window component are included with that component (see L10:C2(b) and L20:C2(b))
N.13.1.b	Architraves, cover fillets and the like	P20:2	Cover fillets, stops, trims, beads, nosings and the like. Architraves included in P20:1
N.13.1.c	Stops		Included above in P20:2
N.13.1.d	Glazing beads and the like		Included above in P20:2
N.13.1.e	Isolated shelves, worktops, seats and the like	P20:3	Isolated shelves and worktops
N.13.1.f	Window boards, nosings, bed moulds and the like	P20:4	Window boards Nosings and bed moulds - see P20:2 above

SMM6		SMM7	
Clause	Heading	Clause	Heading/Comment
N.13.1.g	Handrails	P20:7	Isolated handrails and grab rails
		P20:M4	Associated handrails are measured in Work Sections L30 and L31
N.13.2	Backboards, plinth blocks and the like	P20:9	No change
N.13.3	Built up members so described	P20:1-9.*.*.1	No change
N.13.4	Components which are tongued so described	P20:1-9.*.*.2	No change
N.13.5	Applied coverings	–	Not specifically mentioned
N.14	Labours on unframed second fixings	P20:1-9	Members to be described stating the number of different cross section shapes. Stopped labours to be so described. Curved work to be so decribed stating the radii. Notches etc. are not specifically mentioned
		–	

SMM6		SMM7	
Clause	Heading	Clause	Heading/Comment
N.15	Sheet linings and casings	K11 $	<u>Rigid sheet flooring/ sheathing/linings/ casings</u>
		K12 $	<u>Under purlin/inside rail panel linings</u>
		K13	<u>Rigid sheet fire linings/panelling</u>
		K20 $	<u>Timber board flooring/ sheathing/linings/ casings</u>
		K21 $	<u>Timber narrow strip flooring/linings</u>
			Note: References below to K11 apply equally to K12, K13, K20 and K21
			$ See also under SMM6N.4 - Boardings and flooring where the same SMM7 Work Sections are applicable to "first fixings"
N.15.1 (N.1)	Particulars, type and quality of materials	K11:S1-13	Reworded but very similar particulars required to be stated with the addition of:
			S6. Fire retardant treatments
			S7. Details of cover and jointing strips and cover mouldings
			S10. Matching grain and colour
			S11. Fixing through vulnerable materials
N.15.2	Sloping work	K11:1-7.*.*.2	Unchanged but measured in accordance with definition rule D6
N.15.3	Deduction for voids	K11:M1	No change

SMM6		SMM7	
Clause	Heading	Clause	Heading/Comment
N.15.4	Work to walls and ceilings	K11:1 and K11:3	Unchanged except widths not exceeding 300mm no longer required to be stated in 100mm stages
N.15.5	Work to isolated columns, reveals of openings, pipe casings and the like	K11:6 K11:7	Now to include for isolated beams and to be measured by girth in stages of 600mm measured on the external face
N.15.6	Work to ceilings and beams over 3.50m above floor	K11:M2	No change
N.15.7	Applied finishes to edges	K11:9 K11:D9	Finished angles Defined as angles where the decorative veneer or facing is returned or on panelling where angles are other than butt jointed
N.16	Labours on sheet linings and casings		
N.16.1	Labours	K11:C1	All labours deemed to be included except for holes for pipes, standards and the like and any required to be stated as supplementary information
N.16.2	Raking cutting, curved cutting and scribing	K11:C1	Deemed to be included
N.16.3	Rounded and coved angles	K11:9	Measured for internal and external angles on items dealt with in definition Rule D9
N.16.4	Notches etc.	K11:C1	Deemed to be included

SMM6		SMM7	
Clause	Heading	Clause	Heading/Comment
N.16.5	Forming openings	K11:C1	Labours in forming openings are deemed to be included
N.16.6	Forming access panels	K11:12	No change

SMM6		SMM7	
Clause	Heading	Clause	Heading/Comment
N.17	Composite items Generally		The SMM6 section of Composite items of Woodwork has been divided into a number of Work Sections in SMM7 as follows:
		G20	Structural/ Carcassing metal/ timber
		K32	Framed panel cubicle partitions
		L10	Timber windows/ rooflights/screens and louvres
		L12	Plastics windows/ rooflights/screens and louvres
		L20	Timber doors/ shutters/hatches
		L22	Plastics/Rubber doors/shutters/ hatches
		L30	Timber stairs/ walkways/balustrades
		N10	General fixtures/ furnishings/ equipment
		N11	Domestic kitchen fittings
		N20-23	Special purpose fixtures/furnishings/ equipment
		P10	Sundry insulation/ proofing work/fire stops
		Q50	Site/Street furniture/equipment
N.18	Trussed rafters, roof trusses and the like	G20:1-3	Trusses, trussed rafters and trussed beams No change in method of measurement Refer to SMM6 N.2 above

SMM6		SMM7	
Clause	Heading	Clause	Heading/Comment
		L20	**Timber doors/ shutters/hatches**
		L21	**plastics/Rubber doors/ shutters/hatches**
			Note: References below to L20 apply equally to L21
N.19	Doors	L20:1	Doors Generally unchanged but more comprehensive rules included together with classifications for:-
		L20:3	Sliding/folding partitions
		L20:4	Hatches Note: other classifications are included but these are more applicable to metal see SMM6 trade Q
		L20:M4	Doors where supplied with the associated frames or linings are measured as composite items under General Rule 9.1
N.20	Door frame and lining sets	L20:7	Door frames and door lining sets Generally unchanged but rules now included for
		L20:7.6	Composite sets which would include door, architraves, trims and the like, ironmongery, finishes off site, glazing etc.
		L20:8-10	Rules included for bedding and pointing previously covered in G.43 of SMM6

SMM6		SMM7	
Clause	Heading	Clause	Heading/Comment
N.21 and	Casements and frames, window surrounds, sash windows, lantern lights and skylight sets	L10 L12	**Timber windows/ rooflights/screens/ louvres** **Plastics windows/ rooflights/screens/ louvres**
N.22	Screens and borrowed lights	L12	Note: References below to L10 apply equally to L12
			Generally unchanged but more comprehensive rules included together with classifications for:-
		L10:2	Window shutters
		L10:3	Sun shields
		L10:6	Shop fronts
		L10:7	Louvres and frames
		L10:M1	Standard sections are to be identified
		L10:8, L10:9 and L10:10	Rules included for bedding and pointing previously covered in G.43 of SMM6
		L30	**Timber stairs/ walkways/balustrades**
N.23	Staircases and short flights of steps	L30.1	Generally unchanged but more comprehensive rules included
		L30:3 L30:D2	Associated handrails - these are defined as handrails of a material different from the balustrade with which they are associated Isolated handrails are measured in Section P20 refer to SMM6N.13.1g above

SMM6		SMM7	
Clause	Heading	Clause	Heading/Comment
N.24	Balustrades	L30:2	Isolated balustrades - unchanged
		L30:4	Ramps, wreaths, bends etc. measured extra over
N.25	Sundries	K32	**Framed panel cubicle partitions**
-	No specific section in SMM6 for cubicle partitions but are often measured in this position	K32:1	Cubicle partitions; set to be enumerated with dimensioned diagrams
		K32:2	Trim measured in linear metres and referred to separate items as definition rule D2

SMM6		SMM7	
Clause	Heading	Clause	Heading/Comment
N.26	Supply of fittings	N10	**General fixtures/ furnishings/ equipment**
N.27	Fixing fittings	N11, N20	**Domestic kitchen fittings**
		N21, N22 and N23	**Special purpose fixtures/ furnishings/ equipment**
			Note: References below to N10 apply equally to N11, N20, N21, N22 and N23
		N10:1	Fixtures furnishings and equipment not associated with services to be enumerated with a component drawing reference or a dimensioned diagram
		N10:2	Dealt with in SMM6 trade V
		N10:3	Carving and sculpting - new to SMM7 with regard to timber items to be fully described and enumerated
		N10:4-5	Dealt with in SMM6 trades R and S
		N10:6	Fixtures, furnishings, equipment, fittings and appliances provided by the Employer to be enumerated stating the type, size and method of fixing Appendix A to SMM7 lists the items which should be included within the above Work Sections

SMM6		SMM7	
Clause	Heading	Clause	Heading/Comment
	Sundries		
N.28	Plugging	–	Method of fixing is to be given with the item being fixed
N.29	Holes in timber		Holes for bolts and the like in timber are deemed to be included but holes for pipes standards and the like are measurable with the particular Work Sections - refer to G20:20-28.C3 and H20:10. This also applies to H21, K11, K12, K13, K20 and K21
N.30	Insulating materials	P10	**Sundry insulation/ proofing work/fire stops**
N.30.1	Insulating materials measured in square metres	P10:1-4	Rules have been expanded but requirements are little changed
		P10:D1-4	Definitions are given for:- D1 - Type of work included and types of material D2 - Horizontal includes the upper surface of any sloping structure not exceeding 45 degrees from horizontal D3 - Vertical includes the upper surface of any sloping structure exceeding 45 degrees from horizontal

SMM6		SMM7	
Clause	Heading	Clause	Heading/Comment
			D4 - Soffit includes the underside of any horizontal or sloping structure
N.30.2	Raking and curved cutting	P10:C1	All cutting is deemed to be included
N.31	Metalwork	G20:20-28	All items of straps, hangers, shoes, nail plates, metal connectors, bolts, rod bracing, wire bracing etc. are enumerated and all labours, fixing
		G20:C3-4	and accessories are deemed to be included
		G20:D13	Bolts are measured overall the head and includes heads, nuts and washers
	<u>Ironmongery</u>	P21	<u>Ironmongery</u>
		N15	<u>Signs/Notices</u>
N.32	Generally	P21:1	No change
			Appendix A to SMM7 lists the items which should be included within Work Sections P21 and N15
			Items of ironmongery supplied with a component (e.g. window furniture) are deemed to be included with that component
	<u>Protection</u>		
N.33	Protecting the work		Dealt with in preliminaries under "Employer's requirements" and
		A34:1.6	
		A42:1.11	"Contractor's general cost items"

SMM6		SMM7	
Clause	Heading	Clause	Heading/Comment
P	**STRUCTURAL STEELWORK**	G	**STRUCTURAL/CARCASSING METAL/TIMBER**
		G10	**Structural Steel framing**
			SMM7 has two new sections for structural metal:
		G11	**Structural aluminium framing**
		G12	**Isolated structural metal members**
			References below to G10 apply equally to G11 and G12 unless specifically noted
	Generally		
P.1	Information		
P.1.1	General description of work where not on drawings	–	Option of providing a general description where work is not shown on location drawings has been omitted. The work is to be shown on location drawings
P.1.2	Information to be provided	G10:P1	No change
P.2	Plant	A43	Dealt with in preliminaries
P.3	Classification of work	G10:1-5	Items of structural work now divided into classifications of framing, fabrication; framing, erection; permanent formwork; cold rolled purlins and cladding rails; isolated structural members.

SMM6		SMM7	
Clause	Heading	Clause	Heading/Comment
		G10:3	Unfabricated work would be measured as "Isolated structural member; plain member; use stated" Permanent formwork - new section for SMM7 deals with steelwork which is structurally integral with the framing
		G10:4	Cold rolled purlins and cladding rails - new section for SMM7
	Steelwork		
P.4.1	Generally	-	No change in the general unit of measurement i.e. tonne
P.4.2	Particulars to be given	G10:S1-3	Only type and grade of material and details of welding tests and X-rays and performance tests required to be stated
P.4.3	Particulars to be given in description	G10:1	Only the mass of the units needs to be stated and is to include all components and fittings (except where of a different type and grade of material). Fittings are only measured separately where of a different type of grade of material.
		G10:M1	
		G10:M2	
		G10:1.1-8. 0.1-4	Tapered and castellated members now need to be stated in addition to curved or cambered

168

SMM6		SMM7	
Clause	Heading	Clause	Heading/Comment
P.5	Designation of members	–	No longer a requirement to state sizes of members or components
P.6	Weights of steelwork	G10:M1-3	No change to the rules relating to computation of weight except the mass of metal removed to form notches and holes each > 0.10m2 shall now be deducted from the overall weight.
		G10:M4	No allowance in calculation of mass is made for weld fillets, black bolts, nuts, washers, rivets and protective coatings
P.7	Lengths of members	–	No longer a requirement to state lengths
P.8	Fittings, connections, fixings and anchorages		
P.8.1	Fittings	G10:M1-2	Fittings now included with main member unless of a different type and grade of material in which case they are measured
		G10:1.10	under G10:1.10 as fittings in tonnes
P.8.2 and P.8.4	Connections and anchorages	G10:1.11 and G10:1:12	Holding down bolts or assemblies, special bolts and fasteners only are described and enumerated
P.8.3	Shop and site black bolts	G10:C1	Shop and site black bolts now deemed to be included and no allowance is made for their mass

SMM6		SMM7	
Clause	Heading	Clause	Heading/Comment
P.9	Painting and other surface treatments	G10:7-8	All surface preparation and treatments (including galvanizing) to be measured in square metres stating type and details of application and the time when it is to be applied
		G10:S4	
		G11:9	Localised protective coatings (to structural aluminium framing only) are also measured in square metres
P.10	Erection	G10:2.1 and G10:2.2	Trial erection and permanent erection on site shall be stated in tonnes and is deemed to include all operations subsequent to fabrication
	Protection		
P.11	Protecting the work		Dealt with in preliminaries under "Employer's requirements" and "Contractor's general cost items"
		A34:1.6	
		A42:1.11	

There is no comparable item in SMM6 for the following:-

| | | G10:6 | Filling hollow sections |
| | | | To be given as items with full details stated |

SMM6		SMM7	
Clause	Heading	Clause	Heading/Comment
Q	**METALWORK**	L N P	<u>WINDOWS/DOORS/STAIRS</u> <u>FURNITURE/EQUIPMENT</u> <u>BUILDING FABRIC SUNDRIES</u>
		L11	<u>Metal Windows/ rooflights/screens/ louvres</u>
		L12	<u>Plastics windows/ rooflights/screens/ louvres</u>
			Rules now included for these items which if plastics coated timber were measured under Section N and if plastics coated metal under Section Q of SMM6
			Note: References below to L11 apply equally to L12
	<u>Generally</u>		
Q.1	Information	L11:P1	Information to be shown on location drawings
Q.1.1	Particulars to be given	L11:S1 & S8	No change
Q.1.2	Plates, bars, sections and tubes deemed to be formed by casting, extruding, etc.		No longer a section for plates, bars, sections and tubes. It would seem that such items would be dealt with in an appropriate Work Section.
Q.1.3	Where no method of jointing or form of fabrication given it shall be at the Contractor's discretion	L11:S7	Method of jointing or form of construction to be given

SMM6		SMM7	
Clause	Heading	Clause	Heading/Comment
Q.1.4	Finish of fixings shall match the material being fixed unless otherwise stated	-	Not mentioned
Q.1.5	Priming or painting off site and proprietary finishes or surface treatments so described	L11:S2-4	No change
Q.1.6	Curved items so described	-	Not mentioned
Q.1.7	Lugs so described	L11:S8	Include within method of fixing
	Composite items		
Q.2	Generally		
Q.3	Windows and doors door		
	Note: Clauses relating solely to doors or other items are dealt with separately		Note: Details of L21 - Metal doors/ shutters/hatches follow this Work Section
Q.3.1	Windows complete with frames, mullions, transoms, hinges and fastenings	L11:1	Windows and window frames generally unchanged but classifications also included for:-
		L11:2	Window shuttters
		L11:3	Sun shields
		L11:5	Screens, borrowed lights and frames
		L11:6	Shop fronts
		L11:7	Louvres and frames
Q.3.2	Opening gear	L11:C1(f)	Deemed to be included where supplied with the component

SMM6		SMM7	
Clause	Heading	Clause	Heading/Comment
Q.4	Rooflights, laylights and pavement lights		
Q.4.1	Rooflights and laylights	L11:4	Rooflights, skylights, roof windows and frames. Additional components otherwise unchanged
Q.4.2	Pavement lights		Not specifically mentioned
		L11:8 9 & 10	Rules included for bedding and pointing previously covered in G43 of SMM6
Q.3	Windows and doors	L21	**Metal doors/shutters/ hatches**
		L22	**Plastics/rubber doors/ shutters/ hatches**
			Note: References below to L21 apply equally to L22
Q.3.1	Doors complete with frames, hinges and fastenings	L21:1	Doors Generally unchanged but classifications also included for:-
		L21:2	Rolling shutters and collapsible gates
		L21:3	Sliding/folding partitions
		L21:4	Hatches
Q.3.3	Door frames except those forming an integral part of a door	L21:7	Door frames and door lining sets. Rules are now the same for those in timber see Section L20
Q.3.4	Fireproof and Strongroom doors	L21:5	Strongroom doors - no specific mention of fireproof doors
		L21:6	Grilles

SMM6		SMM7	
Clause	Heading	Clause	Heading/Comment
		L21:8 9 & 10	Rules included for bedding and pointing previously covered in G.43 of SMM6
		L31	**Metal stairs/ walkways/ balustrades**
Q.5	Balustrades and staircases		
Q.5.1 & 4	Balustrades and railings	L31:2	Isolated balustrades
		L31:4	Ramps wreaths bends opening portions etc. measured extra over. Generally unchanged
Q.5.2	Staircases	L31:1	No requirement now to give the weight
Q.5.3 & Q.5.5 & Q.5.6	Handrails and core rails Handrail brackets Isolated balustrades and newels	L31:2-4	Isolated balustrades, Associated handrails both given in linear metres. Labours on same are measured extra over and enumerated
		P20 P20:7	Isolated handrails and grab rails are measured in linear metres in "Building fabric sundries"
Q.6	Sundries		
Q.6.1	Duct covers	P20:6 P20:C1	No change except labours deemed to be included
Q.6.2	Gates, collapsible gates and revolving shutters	L21:2	Collapsible gates are measured as above.
		Q40:5	Gates are only mentioned with regard to fencing and are to be enumerated with height, width and type stated.

SMM6		SMM7	
Clause	Heading	Clause	Heading/Comment
		Q40:C5	Gates are deemed to include gate stops, gate catches and independent gate stays and their associated works
		Q40:3	Independent gate posts are enumerated stating type, height and depth below ground and are deemed to include
		Q40:C1	post holes and
		Q40:C2	slamming stops and hanging fillets. Internal gates (i.e. security gates) would we consider be
		L21:1	measured as doors with a fully dimensioned diagram and enumerated giving the approximate weight
Q.6.3	Cloakroom fittings, cycle-racks, storage racks and the like	N10:1 N12:1 N12:4 or N20-23:1	Cloakroom fittings, catering equipment, etc. will be enumerated with dimensioned description, dimensioned diagram or component drawing as appropriate (Refer to Appendix A of SMM7)
Q.6.4	Grilles and gratings	L21:6	No change in measurement rules
Q.6.5	Ladders	N10:1	Measure as furniture/equipment
Q.6.6	Surface boxes and inspection covers		These items are not separately identified but could be measured under P20

SMM6		SMM7	
Clause	Heading	Clause	Heading/Comment
Q.6.7	Sectional tanks		The only reference we can find to tanks is in the Common Arrangement definition for Y21 - Water tanks/cisterns to be measured as Equipment
		Y21:1	
Q.6.8	Structural metalwork		Structural metalwork is to be measured under:-
		G10	**Structural steel framing**
		G11	**Structural aluminium framing**
		G12	**Isolated structural metal members** as appropriate
	Plates, bars, etc.		
Q.7	Floor plates		
Q.7.1	Floor plates, plain, chequered or perforated		The only reference we can find to chequer plate flooring and the like is in the Common Arrangement definition for items excluded from G30 and cross referenced to L31 - Metal stairs/walkways/balustrades. Under L31 composite items are enumerated with a dimensioned description or component drawing
Q.7.2	Frames, arch bars, bearers etc.	F30:12-14	Arch-bars and the like for air-bricks, ventilating gratings, soot doors etc., are measured with the component
		F30:16	Proprietary metal lintels are to be measured in Masonry Accessories

SMM6		SMM7	
Clause	Heading	Clause	Heading/Comment
Q.7.3	Labours		Measure in accordance with the relevant rules in the appropriate Work Section
Q.7.4	Skirtings and similar trims	P20:1-2	Building fabric sundries. No change except labours are deemed included
Q.7.5	Straps, collars, hangers, brackets and corbels	P20: N10:	Not separately identified but would be measured either under "Building fabric sundries" sundry items or "Furniture and Equipment" as appropriate
Q.7.6	Mat-frames	N10:1	Mat-frames are enumerated as "Furniture and Equipment. Refer to Appendix A to SMM7
	Sheet metal, wiremesh and expanded metal		
Q.8	Coverings	M30:1-8	Coverings now classified into walls, ceilings, beams, etc. Widths not exceeding 300mm are so described and actual width need not be stated. No mention is made of cutting or of welted or beaded edges.

SMM6		SMM7	
Clause	Heading	Clause	Heading/Comment
	Holes, bolts, screws and rivets		
Q.9	Generally		These items are only measured where required by the rules appropriate to the component being measured
	Protection		
Q.10	Protecting the work	A34:1.6 A42:1.11	Dealt with in preliminaries under "Employer's requirements" and "Contractor's general cost items"

SMM6		SMM7	
Clause	Heading	Clause	Heading/Comment
R	<u>PLUMBING AND MECHANICAL ENGINEERING INSTALLATIONS</u>		
			Plumbing and Mechanical Engineering Services in SMM7 have been separated into five Work Groups as follows:-
		R	<u>DISPOSAL SYSTEMS</u>
		S	<u>PIPED SUPPLY SYSTEMS</u>
		T	<u>MECHANICAL HEATING/ COOLING/REFRIGERATION SYSTEMS</u>
		U	<u>VENTILATION/AIR CONDITIONING SYSTEMS</u>
		X	<u>TRANSPORT SYSTEMS</u>
			Work Group R has its own rules for part of the work group, namely for rainwater and drainage above and below ground. All other Work Sections in R and all of the other Work Groups (S, T, U and X) refer to Work Group Y for their relative rules
			Work Group Y contains the measurement rules for Mechanical and Electrical Services
			The comparison with SMM6 below has been dealt with the above breakdowns of work sections in mind

SMM6		SMM7	
Clause	Heading	Clause	Heading/Comment
			Work sections R12 and R13 "Drainage below ground" and "Land drainage" respectively are compared with SMM6 trade W later in this book
			Comparison of R10 and R11 follows immediately and comparison with work section Y is then dealt with separately SMM6 clause references R.1-41 therefore appear more than once herein

SMM6		SMM7	
Clause	Heading	Clause	Heading/Comment
R	<u>PLUMBING ETC</u>	R	<u>DISPOSAL SYSTEMS</u>
R.4.1.a	<u>RAINWATER INSTALLATION</u>	R10	<u>Rainwater Pipework/ Gutters</u>
R.4.1.b	<u>SANITARY INSTALLATION</u>	R11	<u>Foul Drainage Above Ground</u>
	<u>Generally</u>		NOTE: References to R10 below apply equally to R11
R.1	Information	R10:P1	Information - reworded with more details in measurement rules, coverage rules and supplementary information. See also General Rules Clause 4.5
R.2	Plant	A43	Dealt with in Preliminaries
R.3	Generally		
R.3.1	Regulations tests etc.	R10:S1-4	No change
R.3.2	Assembling and jointing component parts etc.	R10:C1	No change - see General Rules Clause 9.1
R.3.3	Patterns moulds templates etc.	R10:C2	No change
R.3.4	Surface treatments off site	R10:D1 & S5-6	No change except for insulation which is measured under Section Y50 and M60
R.3.5	Method of fixing	R10:1.*.*.1	No change
R.3.6	Backgrounds	R10:1.*.*.1 & R10:10.*.*.1	Backgrounds now classified in groups. See General Rules Clause 8

SMM6		SMM7	
Clause	Heading	Clause	Heading/Comment
R.3.7	Temporary work	-	See Additional Rules R10-13. Drainage - work to existing buildings. Clause 7. Now required to be stated if fabricated prior to installation
R.3.8	Connections and breaking into existing	R10:D2 & Additional Rules Clause 7	Now required to state location of work, preparation of ends of existing work and isolation of the system giving limitations on shut down
R.4	Classification of Work	R10 & R11	Measured under the sections stated. See also Code of Procedure for Measurement Sections R10 & R11 Generally. In addition each project to be treated on its own merits.
R.4.1.b	Sanitary Installation (where applicable to laboratory and industrial waste)	R14	**LABORATORY/ INDUSTRIAL WASTE DRAINAGE** Measured in accordance with Section Y
R.4.1. c-p & s	Mechanical Installations		**INSTALLATIONS NOT COVERED BY DISPOSAL SYSTEMS** Measured in accordance with Section S, T, U and Y
R.4.1.q	Vacuum Installation	R30	**CENTRALISED VACUUM CLEANING** Measured in accordance with Sections S and Y

SMM6		SMM7	
Clause	Heading	Clause	Heading/Comment
R.4.1.r	Refuse Disposal Installation	R31	**REFUSE CHUTES** Measured in accordance with Section Y
		R32	**COMPACTORS/MACERATORS** New Section - Measured in accordance with Section Y
R.4.1.u	Special Equipment	R33	**INCINERATION PLANT** Measured in accordance with Section Y
R.5	Location of work	R10:P1	To be as shown on drawings. See also General Rules Clause 7.1
	Gutterwork		
R.6	Generally - Purpose made gutters and fittings	R10:10	Not specifically mentioned but would be fully described or accompanied by component drawings
R.7	Gutters and Fittings		
R.7.1	Gutters	R10:C9	Joints in running length now deemed included
R.7.2	Special joints in running length	R10:D3	No change
R.7.3	Fittings	R10:M7	Reducing fittings now measured extra over largest gutter
R.7.4	Materials for jointing	R10:C1	No change. See General Rules Clause 4.6(b)

SMM6		SMM7	
Clause	Heading	Clause	Heading/Comment
R.8	Gutter supports		
R.8.1	Standard supports	R10:10.*.1	No change
R.8.2	Special supports	P31:30	Fabricated supports to be enumerated
& R.8.3	Fixing supports	R10:10.*.1 & R10:11.1.1.1	giving full details, method of fixing and background in accordance with General Rule 8
	Pipework		
R.9	Generally		
R.9.1	Purpose made pipes and fittings	R10:1	Not specifically mentioned but would be fully described or accompanied by component drawings
R.9.2	Surface treatments	R10:D1 & S5-6	As previous comments above (SMM6-R.3.4)
R.9.3	Pipework temporarily fixed for chromium plating or other special finishing	R10:S5-6	Temporary fixing not mentioned. Off site finishes or surface treatments to state if before or after fabrication
R.10	Pipes and fittings		
R.10.1	Pipes	R10:C3-4	Joints in the running length and joints necessary solely for erection purposes now deemed to be included
R.10.2	Special joints in the running length	R10:D2	No change
R.10.3	Flexible and extensible pipes	R10:M2	No change
R.10.4	Pipes laid in ducts, trenches etc.	R10:1.*.*.2-5	No change

SMM6		SMM7	
Clause	Heading	Clause	Heading/Comment
R.10.5	Flow and return header pipes	-	Not applicable to R10-13 - refer to Work Section Y10
R.10.6	Fittings	R10:2	No change
R.10.7	Expansion loops	-	Not applicable to R10-13 - refer to Work Section Y10
R.10.8	Expansion compensators	R10:D2	Measure extra over pipe as special connection
R.10.9	Flanges and unions	R10:C4	No change
R.10.10	Materials for jointing	R10:C1	No change
R.11	Sockets etc.	R10:3-5	No change
R.12	Special connections etc.		
R.12.1	Special connections	R10:D2	Now measured as extra over the pipes in which they occur
R.12.2	Isolated joints in the running length	R10:D2	No change
R.12.3	Connecting pipes to boilers etc.	R10:D2	Now measured as extra over the pipes in which they occur
R.13	Labours	R10:C5	Now all deemed to be included except made bends
R.14	Sundries		
R.14.1-5	Rainwater heads, flashing-plates etc.	R10:6	No change except number of supports to be stated and fixing to be stated as for pipes. Backgrounds as previously stated

SMM6		SMM7	
Clause	Heading	Clause	Heading/Comment
R.14.6	Pipe sleeves	R10:8	Length now given in stages of 300mm. State method of fixing or if to be handed to others for fixing
R.14.7	Wall, floor or ceiling plates	R10:9	No change
R.15	Pipework supports		
R.15.1	Standard pipe supports	R10:1.*.1	Included with pipelines as SMM6 except supports which differ shall be enumerated. Supports for more than one service shall be enumerated in accordance with Section P31 Clause 30
		R10:7 & M5	
R.15.2 & 4	Non-standard pipe supports	R10:7 & M5 P31:30	No change except fabricated supports and supports for more than one service to be enumerated giving full details and method of fixing and background in accordance with General rule 8. Measurement Rules R10:M7 relates this work to Work Section P30 - we believe this is incorrect and that it should be referred to Work Section P31
R.15.3	Lined or insulated pipe supports	R10:7.*.*.1	No change
R.15.5	Pipe anchors and guides	-	Not specifically mentioned - assume measure in accordance with Clause 7 or Section P31 Clause 30
R.15.6	Fixing of supports etc.	R10:1 & 7	As previously described for pipes

SMM6		SMM7	
Clause	Heading	Clause	Heading/Comment
R.17-21	<u>Ductwork</u>	-	Refer to the comparison with Work Section Y below
	Equipment and ancillaries		
R.22.1-4	Sanitary appliances and similar items		To be measured under Work Group N - FURNITURE/EQUIPMENT Work Section N13: Sanitary appliances/fittings
		N12:4-6 N13:4-6	Fittings, equipment and appliances associated with services to be enumerated with full details
R.22.5	Platework, fuel hoppers and supporting steelwork	N13:4	Ancillaries and supports provided with the fittings to be included in the description of the fittings
R.23 & R.24	Heater elements and casings, Anti-vibration and sound insulation	-	Refer to the comparison with Work Section Y below
R.25	Equipment supports	N13:4-5	Either included in description of fitting or enumerated separately
R.26	Jointing	N13:C1-2	All jointing of equipment and ancillaries is deemed to be included
R.27	Chimneys	-	Refer to the comparison with Work Section Y below
R.28	Loose ancillaries	N13:4.*. *.1	Ancillaries provided with the equipment to be included in the description of the equipment

SMM6		SMM7	
Clause	Heading	Clause	Heading/Comment
		N13:5	Ancillaries not provided with the equipment to be enumerated separately
R.29	Identification plates	R10:13	Now only required to be measured separately from items they identify
R.30-36	<u>Insulation</u>	-	Now measured in accordance with Section Y50
	<u>Sundries</u>		
R.37	Generally		
R.37.1	Draining and refilling existing systems	-	See Additional Rules R10-13. Drainage - work to existing buildings-draining down part or whole of systems
R.37.2	Marking positions of holes etc.	R10:12.*.*.1	Details now required to be stated for items formed during construction
R.37.4	Temporarily operating installations	R10:15	No change
R.37.5	Testing etc	R10:14.1.1 & C11	Now required to state instruction of personnel in operation of completed installation. Provision of water and other supplies now deemed to be included. Refer to Y51:C1 where a similar coverage rule appears but where the words "fuel, gas, electricity" have been included.

SMM6		SMM7	
Clause	Heading	Clause	Heading/Comment
			Clarification will be needed if "other supplies" in coverage rule R10:C11 are to be deemed to include "fuel, gas, electricity"
R.37.6	Preparing plans etc.	R10:16 & D4	Drawings are now to include builders work. Now required to give details of negatives, prints and microfilms, binding into sets and names of recipients
		R10:17	New item - Operating and maintenance manuals to be given as an item
	Builders Work		
R.38	Generally		Now generally measured under "Building Fabric Sundries" - Work Section P
		P30	<u>Trenches/Pipeways/ Pits for buried engineering services</u>
		P31	<u>Holes/Chases/Covers/ Supports for services</u>
		P30:M2	Builders work in connection with plumbing, mechanical and electrical installations are each to be identified under an appropriate heading
R.39	Electrical Work		See later comparison

SMM6		SMM7	
Clause	Heading	Clause	Heading/Comment
R.40	Particularly		
R.40.1	Excavating trenches for pipes	P30:1 P30:C1	Trenches for services not exceeding 200mm nominal size are grouped together; those for services over 200mm nominal size are measured separately stating the nominal size of the service. Average depth is given in 250mm stages with other specific requirements stated. Excavating trenches are deemed to include earthwork support, consolidation of trench bottoms, trimming excavations, special protection of services (specified protection is to be given as supplementary information (S2) where required), backfilling with and compaction of excavated materials and disposal of surplus excavated materials
R.40.2	Inspection chambers	P30:9 or P30:16-18	Chambers are either measured in accordance with the rules for manholes (R12) or enumerated giving details of type, size, covers, bedding and jointing
R.40.3	Bedding and ponting component	-	Given in the description of the component

SMM6		SMM7	
Clause	Heading	Clause	Heading/Comment
R.40.4	Refractory lining	–	Dealt with in Work Group F (SMM6 trade G)
R.40.5	Cutting and pinning ends of supports	P31:24-25	No change
R.40.6	Cutting away for and making good	P31:19-22 P31:32-35	Measure in detail Lifting and replacing flooring in existing work and chases are measured in detail
R.40.7	Pylons, poles, wall brackets, pole-brackets, pole stays and the like	P31:30	No change
R.40.8	Boring or excavating holes in the ground for poles and stays	P31:30	Now to be given in the description of the support component
R.40.9	Catenary cables	P31:31	No change
	Protection		
R.41	Protecting the work	 A34:1.6 A42:1.11	Dealt with in preliminaries under "Employer's requirements" and "Contractor's general cost items"

SMM6		SMM7	
Clause	Heading	Clause	Heading/Comment
R	**PLUMBING AND MECHANICAL ENGINEERING INSTALLATIONS**	Y	**MECHANICAL AND ELECTRICAL SERVICES MEASUREMENT**
			There are 37 Work Sections in Work Group Y. These have not been listed separately but are referred to by number below (i.e. Y10:). The headings of each Work Section are given in the contents (SMM7) to this chapter.
			Note: References below to Y10 apply equally to Y11.
			References below to Y20 apply equally to Y21-25, Y40-46, Y52 and Y53.
			References below to Y30 apply equally to Y31.
			References below to Y51 apply equally to Y54 and Y59.
R.1	Information	P1	Information showing scope, extent and location of the work
R.2	Plant	A43	Dealt with in preliminaries
R.3	Generally		
R.3.1-6	Particulars, assembly, patterns, fixing etc.	Y10:S1-6 Y20:S1-7 Y30:S1-6 Y50:S1-5	Generally unchanged with some different wording and particular requirements for individual Work Sections

SMM6		SMM7	
Clause	Heading	Clause	Heading/Comment
R.3.7 R.3.8	Temporary work Connections to existing installations	)))) -	Refer to Additional rules - work to existing building
R.4	Classification of work	-	The classifications have been amended and the Surveyor should now refer to the list of Work Sections for Work Group Y and to the Common Arrangement for the type of work which is to be included within each category. Rainwater pipework/gutters; Foul drainage above ground; Drainage below ground and Land drainage are dealt with separately in Work Group R, Work Sections R10, R11, R12 and R13 respectively. Comparison of R10-13 with SMM6 is dealt with separately immediately before this comparison
R.5	Location of work	Y10:M2 Y20:M2 Y30:M2 Y50:M2	Only work in plant rooms is required to be identified separately
R.6-8	**Gutterwork**	-	Refer to the earlier comparison with Work Section R10
	Pipework		
R.9.1	Purpose made pipes and fittings	Y10:1	Not specifically mentioned but would be fully described or accompanied by component drawings

193

SMM6		SMM7	
Clause	Heading	Clause	Heading/Comment
R.9.2	Surface treatments	Y10:S6	Information regarding surface treatments should be included in the description
R.9.3	Pipework temporarily fixed for chromium plating or other special finishing	Y10:S5-6	Temporary fixing not covered. Off site finishes or surface treatments to state if before or after fabrication
R.10	Pipes and fittings	Y10:1	Now explicit between straight, curved (stating radii), flexible or extendable
R.10.1	Pipes	Y10:C3-4	Joints in the running length and joints necessary solely for erection purposes now deemed to be included
R.10.2	Special joints in the running length	Y10:2.2 & D2	No change
R.10.3	Flexible and extensible pipes	Y10:1.3-4	No change
R.10.4	Pipes laid in ducts, trenches etc.	Y10:1.*.*.2-6	No change
R.10.5	Flow and return header pipes	Y10:1.5	No change
R.10.6	Fittings	Y10:2	Generally no change
R.10.7	Expansion loops	Y10:3	Expansion loops
R.10.8	Expansion compensators	Y10:4	Expansion compensators. Both generally as SMM6 but state background and whether in trenches or ducts
R.10.9	Flanges and unions	Y10:C4	Joints necessary solely for erection are deemed to be included

SMM6		SMM7	
Clause	Heading	Clause	Heading/Comment
R.10.10	Materials for jointing	Y10:C1	No change
R.11	Sockets etc.	Y10:5-7	No change
R.12	Special connections etc.	Y10:2.2 Y10:D2	Generally as SMM6 but measured as extra over the pipes in which they occur
R.13	Labours	Y10:2.1	Now all deemed to be included except made bends
R.14	Sundries		
R.14.1-5	Ancillary items	Y10:8	Measure as pipework ancillaries as appropriate
R.14.6	Pipe sleeves	Y10:11	Generally as SMM6 but now separated into lengths not exceeding 300mm and thereafter in 300mm stages
R.14.7	Wall, floor or ceiling plates	Y10:12	No change
R.15	Pipework pipe supports		
R.15.1	Standard pipe supports	Y10:1.1.1	No change
R.15.2	Non-standard pipe supports	Y10:9	No change except fabricated supports and supports carrying more than one service are measured in accordance with Work Section P31
R.15.3	Lined or insulated pipe supports	Y10:9.*.*.1	No change
R.15.4	Supports for more than one pipe etc.	-	Refer to comment above (SMM6-R:15.2)

SMM6		SMM7	
Clause	Heading	Clause	Heading/Comment
R.15.5	Pipe anchors and guides	Y10:10	No change
R.15.6	Method of fixing	Y10:9-10	To be stated in the description
R.16	Connections to public mains etc.	A53:1.1-2	Work by statutory authorities is to be given as a provisional sum and includes work by public companies responsible for statutory work
	Ductwork		
R.17	Ducting and fittings		
R.17.1 or R.21.1	Ducting	Y30:1 Y30:C3	Generally as SMM6 but joints in the running length and stiffeners are now deemed to be included
R.17.2 or R.21.2	Flexible and extensible ducting	Y30:1.5	Flexible ducting now measured in linear metres. Extensible ducting not specifically mentioned
R.17.3 or R.21.3	Curved ducting	Y30:1.2-4	Now to be measured in linear metres
R.17.4 or R.21.4	Lining ducting internally	Y30:2.1 Y30:M4	Similar to SMM6 but now measured as extra over or may be given in the description of the ducting
R.17.5 or R.21.5	Ducting fittings	Y30:2.3	Generally as SMM6 but no mention of change in direction of ducting being described.

SMM6		SMM7	
Clause	Heading	Clause	Heading/Comment
		Y30:M5	Where there is a preponderance of fittings they may be enumerated separately as full cost items
R.17.6 or R.21.6	Providing all necessary for joints etc.	Y30:C1	No change
R.18	Special connections etc.		
R.18.1 or R.21.7	Special connections and joints	Y30:2.2	Now to be measured as extra over the ducting in which they occur
	New items for SMM7	Y30:5	Breaking into existing ducts to be measured as an item and fully described.
		Y30:7	Ducting sleeves through walls, floors and ceilings to be enumerated in length stages of 300mm
R.18.2	Ducting turns and splitters	Y30:3	No change
R.18.3	Nozzle outlets etc.	Y30:2.4-6	Now to be measured as extra over the ducting in which they occur
R.18.4 & R.18.5	Cowls, terminals etc. Weathering aprons etc.	Y30:4.1	Measure as ancillaries with full description
R.19 or R.21.7	Ductwork supports	Y30:6	Generally as SMM6. Fabricated supports and supports carrying more than one service are to be measured in accordance with Work Section P31

SMM6		SMM7	
Clause	Heading	Clause	Heading/Comment
	Equipment and ancillaries	Y20-25, Y40-46, Y52 and Y53	
R.22	Generally	Y20:1-2	Generally as SMM6
		Y10:8	Pipework ancillaries covers the like of draw-off taps, stop valves etc. measured numerically stating type, size etc. and type of pipe
R.23	Heater elements and casings	Y20:3-4	Generally unchanged except edge sealing strips are now deemed to be included. Backing insulation etc. is not mentioned but could be measured extra over
R.24	Anti-vibration and sound insulation	Y20:8 Y20:9	Anti-vibration mountings Anti-vibration material. Both unchanged
R.25	Equipment supports	Y20:6	No change
R.26	Jointing	-	Generally included with item or deemed to be included in jointing ancillaries
R.27	Chimneys	Y20:7	Generally unchanged except cowls and terminals no longer require the number to be stated
R.28	Loose ancillaries	Y20:2 & Y51:2	No change
R.29	Identification plates	Y51:3	No change

SMM6		SMM7	
Clause	Heading	Clause	Heading/Comment
	Insulation		
R.30	Generally	Y50:S1-5 & C1(a)	No change
R.31	Insulation to pipework	Y50:1-2	Unchanged except working around pipe flanges and around fittings (excluding
		Y50:C1	metal clad facing insulants) is now deemed to be included
R.32	Insulation to ductwork	Y50:1.3	Unchanged except working around fittings (excluding metal clad facing insulants) is now deemed to be included
R.33	Insulation to equipment	Y50:1.4 & Y50:2.3	No change
			Note: Insulation with metal clad facing shall be measured extra over around fittings
R.34	Sundry insulation	Y50:3	Unchanged except special protection or finish at openings through wall etc. should be included stating details
R.35 & R.36	**Insulation (alternative)**	-	No alternative given
	Sundries		
R.37	Generally		
R.37.1	Draining and refilling existing systems	-	Refer to Additional Rules - work to existing buildings

SMM6		SMM7	
Clause	Heading	Clause	Heading/Comment
R.37.2	Marking positions of holes etc.	Y51:1	Details now to be stated for items formed during construction
R.37.3	Disconnecting, setting aside etc.	Y20:10	No change
R.37.4	Temporarily operating each installation etc.	Y51:5 Y51:C1	Unchanged except provision of water, fuel, gas, electricity and other supplies is now deemed to be included
R.37.5	Testing etc.	Y51:4 Y51:C2	Unchanged except the provision of test certificates is now deemed to be included
R.37.6	Preparing plans etc.	Y51:6-7	Drawings are now to include builders work. Nos required to give details of negatives, prints and microfilms, binding into sets and names of recipients
R.38-40	**Builders work**	-	Now generally measured under "Building Fabric Sundries" - Work Section P - Refer to Comparison of R10 with SMM6 earlier in this book
	Protection		
R.41	Protecting the work		Dealt with in preliminaries under "Employer's requirements" and "Contractor's general cost items"
		A34:1.6	
		A42:1.11	

NOTE: The Work Sections within SMM7 Work Groups S, T and U are all to be measured in accordance with the rules of Work Group Y and would be compared with SMM6 as commented above

SMM6		SMM7	
Clause	Heading	Clause	Heading/Comment
S	**ELECTRICAL INSTALLATIONS**	Y	**MECHANICAL AND ELECTRICAL SERVICES MEASUREMENT**
			The following Work Sections relate to Electrical installations:
		Y60	**Conduit and cable trunking**
		Y61	**HV/LV cable and wiring**
		Y62	**Busbar trunking**
		Y63	**Support components - cables**
		Y70	**HV switchgear**
		Y71	**LV switchgear and distribution boards**
		Y72	**Contactors and starters**
		Y73	**Luminaires and lamps**
		Y74	**Accessories for electrical services**
		Y80	**Earthing and bonding components**
		Y81	**Testing and commissioning electrical services**
		Y82	**Identification - electrical**
		Y89	**Sundry common electrical items**
		Y92	**Motor drives - electric**
			Note: The following work sections are grouped together. References below to Y60 apply equally to Y63
			References below to Y61 apply equally to Y62 and Y80.
			References below to Y70 apply equally to Y71, Y72 and Y92.

SMM6		SMM7	
Clause	Heading	Clause	Heading/Comment
			References below to Y73 apply equally to Y74.
			References below to Y81 apply equally to Y82 and Y89
	Generally		
S.1	Information	Y60:P1 Y70:P1 Y73:P1 Y81:P1	Location drawings to be provided to show the scope and location of the work.
		Y61:P1	As above but a distribution sheet setting out the number and location of all fittings and accessories together with a location drawing showing the layout of all points is to be given relative to final sub-circuits
S.2	Plant	A43	Dealt with in preliminaries
S.3	Generally		
S.3.1	Particulars to be given	Y60:S1-4 Y61:S1-4 Y70:S1-4 Y73:S1-4	No change
S.3.2	Assembling and jointing together	Y60:C1 Y61:C1 Y70:C1 Y73:C1	No change
S.3.3	Patterns moulds etc.	Y60:C2 Y61:C2 Y70:C2 Y73:C2	No change

SMM6		SMM7	
Clause	Heading	Clause	Heading/Comment
S.3.4	Priming or painting off site, proprietary finishes and surface treatments	Y60:S5-6 Y61:S5-6 Y70:S6-6 Y73:S5-6	Finishes applied on or off site whether before or after fabrication or assembly shall be stated. Decorative finishes are excluded and are measured in accordance with Work Section M60
S.3.5	Method of fixing to be given in description	Various	No change
S.3.6	Nature of backgrounds	Various	To be stated in the description of the work. No classifications of background are given with the specific rules but the identification of various backgrounds is given in General Rule 8. The rules no longer state that any necessary drilling is deemed included. This will need to be made clear
S.3.7	Temporary work	-	Refer to Additional rules - Electrical services - work in existing buildings
S.4	Classification of work		The classifications are as listed at at the beginning of this section
S.5	Location of work		Location of the work is to be shown on location drawings

SMM6		SMM7	
Clause	Heading	Clause	Heading/Comment
	Equipment and control gear		
S.6	Generally	Y70	
		Y70:1-4	Generally as SMM6 separating switchgear, distribution boards, contactors and starters, and motor drives
S.7	Equipment supports	Y70:5	Supports provided with the equipment are to be given in the description of the equipment. Supports not provided are detailed and enumerated
	Fittings and accessories		
S.8	Generally	Y73:1-3	No change except lamps and luminaires separately identified
S.9	Lighting fittings	Y73:2	Luminaires - the length ranges of pendant drops has been amended to not exceeding 1.00m and actual drop if over 1.00m
		Y73:4	Luminaires and lamps provided by the Employer - all as SMM6 except no requirement to give details of ornamental or unusually expensive fittings. It would still be sensible to give this information

SMM6		SMM7	
Clause	Heading	Clause	Heading/Comment
S.10	Accessories	Y73:5	No change. Details of plugs to be provided with socket outlets. Plugs are deemed to include fuses
	Conduit, trunking and cable trays		
S.11	Conduit and fittings		
S.11.1	Conduit	Y61:M6	Conduit shall be measured (where applicable) in all cases except for final circuits in domestic or similar simple installations
		Y63:C3	Saddles and crampets are now deemed to be included with conduit
S.11.2	Special boxes etc.	Y63:2	Generally no change except no mention in SMM7 of flameproof boxes
S.11.3	Flexible conduits etc.	Y63:1.3-4	As SMM6 except earthing tails have now to be stated
S.1.4	Component and special boxes for making connections etc.	Y63:3-4	No change
S.12	Trunking and fittings		
S.12.1	Trunking	Y60:5	Unchanged except straight and curved work (stating radii) to be kept separate
S.12.2	Fittings	Y60:6	Unchanged except details of bushing material to be given

SMM6		SMM7	
Clause	Heading	Clause	Heading/Comment
S.12.3	Connections between trunking and control gear etc.	Y60:7	No change
S.12.4	Pin racks	Y60:5	No longer to be measured separately but are to be included with items of trunking
S.13	Busbar trunking and fittings		
S.13.1	Busbar trunking	Y62:7	No change
S.13.2	Tap-off units, feeder units, fire barriers and the like	Y62:9-11	No change
S.14	Trays and fittings	Y60:8-10	No change
S.15	Trunking and tray supports	Y60:11-12 Y61:12	No change
	Cables		
S.16	Generally		
S.16.1	Circuits and wires	Y60:C3	Draw wires etc. deemed included with conduits
S.16.2	Cables required to be colour coded	Y61:S7	No change
S.17	Cables		
S.17.1	Length of cables	Y61:M2-3	No change
S.17.2	Classifications of cables	Y61:1.1.1-7	No change
S.17.3	Cables in existing conduits		Refer to Additional rules - Electrical services - work to existing buildings

SMM6		SMM7	
Clause	Heading	Clause	Heading/Comment
S.17.4-7	Cable joints, line taps, cable termination glands and conduits boxes used with cable termination glands	Y61:3-5 Y61:2	No change except a new item has been included:- Flexible cable connections shall be enumerated stating size, type, number etc. and separating lengths ≤ 1.00 and thereafter in 1.00m stages when > 1.00m
S.18	Cable supports	Y61:6 P31:30	Cable supports which differ from those given with cables shall be enumerated stating size of cable type and size of support and method of fixing and to which background separating those fixed to surfaces, to conductors and overhead lines and those suspended from catenary cables. Supports for services not provided with the services installation (e.g. pylons, poles etc) shall be measured in "Building fabric sundries"
S.19	Sundries	Y61:C3	Wall, floor and ceiling plates, cables sleeves and connecting tails are all deemed included with cables
S.20	Connection to public mains	A53:1.1-2	Work by statutory authorities is to be given as a provisional sum and

SMM6		SMM7	
Clause	Heading	Clause	Heading/Comment
			includes work by public companies responsible for statutory work
	Final sub-circuits		
S.21	Cable conduits etc. in final sub-circuits	Y61:19	Very similar to to SMM6 as amended
	Earthing		
S.22	Conductors		
S.22.1	Cables and tapes for earthing	Y80:13	As noted previously - refer SMM6.S.17 No change
S.22.2	Conductor connections etc. and junctions	Y80:14-15	No change
S.22.3	Radius bends etc.		Not mentioned
S.22.4-6	Test clamps, earth electrodes and air-termination points	Y80:16-18	No change
	Ancillaries		
S.23	Loose ancillaries	Y81:3	As SMM6 except names of recipients to be stated
S.24	Identification plates	Y70:C3 Y82:4	Plates, discs and and labels for identification purposes are deemed to be included where provided with the equipment. Where not provided with the equipment or control gear they are to be enumerated separately

SMM6		SMM7	
Clause	Heading	Clause	Heading/Comment
	Sundries		
S.25	Generally		
S.25.1	Marking position of holes etc.	Y89:2	No change
S.25.2	General earth bonding	Y80:	Earthing and bonding components are measured in detail in Work Section Y80
S.25.3	Disconnecting etc. for convenience of other trades	Y73:6	No change
S.25.4	Temporarily operating installation	Y81:6	Generally unchanged, provision of electricity and other supplies is covered by provisional sums (A54)
S.25.5	Testing	Y81:5	Testing and commissioning is generally as SMM6 but the provision of instruments is to be given in the description and test certificates are deemed to be included. No mention of stating any special insurance cover required.
		Y81:1	Additional bonding resulting from testing extraneous metal is to be included as a Provisional Sum - this is a new requirement
S.25.6	Preparing plans etc.	Y81:7-8	Preparing drawings very similar to previous requirements except builders work also to be shown. Operating and

SMM6		SMM7	
Clause	Heading	Clause	Heading/Comment
			maintenance schedules now to be provided
	Builders work		
S.26	Generally	P30:M2	Builders work in connection with plumbing, mechanical and electrical installations are each to be identified under an appropriate heading
S.27	Particularly		
S.27.1	Excavating trenches for cables and pipe ducts	P30:1	Trenches for services not exceeding 200mm nominal size are grouped together; those for services over 200mm nominal size are measured separately stating the nominal size of the service. Average depth is given in 250mm stages with other specific requirements stated. Excavating trenches is deemed to include earthwork support, consolidation of trench bottoms, trimming excavations, special protection of services (specified protection is to be given as supplementary information (S2) where required), backfilling with and compaction of excavated material and disposal of surplus excavated materials

210

SMM6		SMM7	
Clause	Heading	Clause	Heading/Comment
	Sand for bedding cables	P30:4	Beds are measured in linear metres stating width and thickness
S.27.2	Cable covers	P30:12	No change
S.27.3	Inspection chambers	P30:9 or P30:16-18	Chambers are either measured in accordance with the rules for manholes (R12) or enumerated giving details of type, size, covers, bedding and jointing
S.27.4	Bedding and pointing components or units of equipment		Given in the description of the component etc.
S.27.5	Cutting and pinning ends of supports	P31:24-25	No change
S.27.6	Cutting away for and making good in new structures	P31:19	No change except making good vulnerable materials has been added
S.27.7	Cutting away for and making good in existing structures	P31:32-35	Measured in detail similar to SMM6
S.27.8	Pylons, poles, hut-posts, wall-brackets, pole-brackets, pole-stays and the like	P31:30	No change
S.27.9 and S.27.10	Boring or excavating holes in the grounds for poles, etc. Excavating pits and forming concrete bases for pylons	P31:30	Now to be given in the description of the support component

SMM6		SMM7	
Clause	Heading	Clause	Heading/Comment
S.27.11	Catenary cables	P31:31	No change
	Protection		
S.28	Protecting the work	A34:1.6 A42.1.11	Dealt with in preliminaries under "Employer's requirements" and "Contractor's general cost items"

SMM6		SMM7	
Clause	Heading	Clause	Heading/Comment
T	<u>FLOOR WALL AND CEILING FINISHINGS</u>	J	<u>WATERPROOFING</u>
		K	<u>LININGS/SHEATHING/DRY PARTITIONING</u>
		M	<u>SURFACE FINISHES</u>
		Q	<u>PAVING/PLANTING/ FENCING/SITE FURNITURE</u>

As stated earlier the authors of SMM7 have departed from the traditional traded format and have tried to align the new document with the requirements of the present day industry into numerous work sections.

The SMM6 floor wall and ceiling finishings trade is one of the trades which has been most affected by this fragmentation into work sections. An easy comparison of the two documents is not possible and we have therefore had to deal with each set of work sections quite separately. The same SMM6 references therefore appear in more than one place below. Also, because the SMM6 clauses are cross referenced to other clauses we have adopted the approach of writing cross references in brackets (e.g. Clause T.13 refers back to clause T.4-10 and therefore when using T.4 in connection with T.13 we have written (T.4))

	<u>Generally</u>		
T.1	Information	P1	The scope and location of the work is to be shown on location drawings; the option of providing a general description of the work has been removed
T.2	Plant	A43	Dealt with in preliminaries

SMM6		SMM7	
Clause	Heading	Clause	Heading/Comment
T.3	Generally		All work sections have their own general rules and these are delat with at the beginning of each work section below
T.4-12	<u>In-situ finishings</u>		The following work sections or parts thereof would previously have been measured under "In-situ finishings"
		J10	<u>Specialist waterproof rendering</u>
		M10	<u>Sand cement/Concrete/ Granolithic screeds/ flooring</u>
		M12	<u>Trowelled bitumen/ resin/rubber-latex flooring</u>
		M20	<u>Plastered/Rendered/ Roughcast coatings</u>
		M21	<u>Insulation with rendered finish</u>
		M22	<u>Sprayed mineral fibre coatings</u>
		M23	<u>Resin bound mineral coatings</u>
		M30	<u>Metal mesh lathing/ Anchored reinforcement for plastered coating</u>
		M41	<u>Terrazzo tiling/ in-situ terrazzo</u>
		Q22	<u>Coated macadam/ Asphalt roads/ pavings</u>
		Q23	<u>Gravel/Hoggin roads/pavings</u>
		Q26	<u>Special surfacings/ pavings for sport</u>
			Note: References below to M10 apply equally to J10, M12, M20 and M23

214

SMM6		SMM7	
Clause	Heading	Clause	Heading/Comment
			Comparison of the rules for M21, M22, M30, M41, Q22, Q23 and Q26 follow this section in sequence
(T.3.1)	Work classified as Internal or External under an appropriate heading	M10:D1	All work is deemed internal unless described as external
(T.3.2)	Work in compartments not exceeding 4.00 square metres on plan	-	No longer a requirement for these areas to be kept separate
(T.3.3)	Overhand work	M10:1-6. *.*.4	No change
(T.3.4)	Grouping of work		No change
(T.3.5)	Patterned work so described	M10:1-6. *.*.1	No change (see also M10:C2)
(T.3.6)	Nature of base	M10:S5	No change
(T.3.7)	Preparatory work	M10:S6	No change
(T.3.8)	Curved work - conical, spherical, convex, concave or to more than one radius kept separate	M10:M5	No change except conical, spherical, convex, concave not mentioned. This information could be given as supplementary information
(T.3.9)	Ceilings and beams over 3.50m above floor. Where walls are over 3.50m high and ceiling finish differs from that on wall details to be given	M10:M4	No change for ceilings and beams. No requirement to give details where work to walls is over 3.50m high above floor and the ceiling finish differs

SMM6		SMM7	
Clause	Heading	Clause	Heading/Comment
(T.3.10)	Work of a repairing nature and to isolated areas not exceeding 1.00 square metres	-	No special requirements. Includes isolated areas with other work of the same material
(T.3.11)	Temporary rules, temporary screeds and templets. Temporary support to face of risers	M10:26	No specific rules Temporary support to face of risers to be measured in linear metres stating height and if undercut
(T.3.12)	No deduction to be made for grounds	M10:M2	No change
T.4	Generally		
T.4.1	Particulars to be given	M10:S1-4	No change except that the thickness is defined as being the "nominal" thickness (D2)
T.4.2	Work to be given in square metres, no deduction made for voids not exceeding 0.50m2. Work not exceeding 300mm to be so described	M10:M2	No change except work not exceeding 300mm is to be measured in linear metres
T.4.3	Lathing, reinforcement and baseboarding given separately	M10:1-6.*.2.* M10:C3	Plasterboard or other rigid sheet lathing is given in the description of the finish. Plasterboard etc. is deemed to include joint reinforcing scrim. Reinforcement is measured separately as an accessory under M10:24 or if metal mesh lathing or

SMM6		SMM7	
Clause	Heading	Clause	Heading/Comment
			anchored reinforcement in Work Section M30 below
T.5.1-4	Walls, ceilings, beams and columns	M10:1-4	No change except as affected by general rules above
T.5.5 and T.5.6	External angles so described Rounded internal and external angles	M10:C4	Internal and external angles and intersections not exceeding 10mm radius are deemed to be included; those between 10 and 100mm radius are measured in linear metres; those over 100mm radius are
		M10:M5	measured as curved work stating the radius
T.5.7	Angle screed, casing and similar beads	M10:24.8	No change
T.6	Mouldings	M10:17-23	No significant changes. A typographical error appears to have occurred in item M10:17-22.*.*.7 where the words "weathered tops" should not appear. This will presumably be corrected by an amendment or subsequent re-print
T.7	Staircase areas		
T.7.1	All work to staircase areas to be given separately	M10:M3	No change except that this clause has been extended to include plant rooms. All work in plant rooms to be given separately

SMM6		SMM7	
Clause	Heading	Clause	Heading/Comment
T.7.2	Work to landings so described	M10:D4	Floors include landings
T.7.3	Treads and risers	M10:7-8	No change except internal and external angles not exceeding 10mm are deemed to be included
T.7.4	Nosings, fair edges, inserts and bands - given as extra over	M10:C6	Fair edges are deemed to be included
		M10:7-12. *.*2 .2	Inserts and bands included in the description of the work
		M10:24.9	Nosings are measured as "Accessories" and not as "Extra over"
T.7.5	Strings and aprons	M10:9-10	No change except ends, angles, ramped and wreathed corners and intersections not exceeding 10mm radius are deemed to be included
T.8	Floors		
T.8.1	Classification	M10:5	Slight change in classification as follows:- .1 Level and to falls only not exceeding 15 degrees from horizontal .2 To falls and crossfalls and to slopes not exceeding 15 degrees from horizontal .3 To slopes over 15 degrees from horizontal

SMM6		SMM7	
Clause	Heading	Clause	Heading/Comment
		M10:6	An additional category of "Roofs" has been included
T.8.2	Work in boiler rooms, machine shops and similar locations to be so described	M10:M3	Work in plant rooms to be so described. This now covers wall and ceiling finishings as well as floors
T.8.3	Floors laid in bays	M10:1-6. *.*.2	No change
T.8.4	Floors laid in one operation with base	M10:1-6. *.*.3	No change
T.9	Channels		
T.9.1	Working finishes into shallow channels - given as extra over	M10:C5	Forming shallow channels is deemed to be included
T.9.2	Linings to channels	M10:12	No change except arrises, coves, ends, angles and intersections and outlets are deemed to be included
T.10	Skirtings and kerbs	M10:13 M10:14 M10:15	Skirtings Kerbs Cappings No change except fair edges, rounded edges, beaded edges, coved junctions, ends, angles and ramps are deemed to be included
T.11	Labours	M10:C1	All are deemed to be included
T.12	Dividing strips	M10:24.7	No change

SMM6		SMM7	
Clause	Heading	Clause	Heading/Comment
T.4-12	<u>In-situ finishings</u> (continued)	M21	<u>Insulation with rendered finish</u>
		M21:M1	Only proprietary construction is measured in this Section
		M21:D1	Work in this Section is deemed to be <u>external</u> unless described as internal
T.4-5	Walls, ceilings and beams	M21:1-3	No change except work not exceeding 300mm is to be measured in linear metres and the work is deemed to include accessories for fixing
T.5.7	Beads	M21:4	No change
T.7.4	Nosing	M21:5	Not measured "extra over" the work on which they occur, otherwise no change
T.12	Dividing strips	M21:6	Expansion strips - no change. Note: Beads, nosings and expansion strips are not called "Accessories" in this Work Section

	SMM6		SMM7	
Clause	Heading	Clause	Heading/Comment	
T.4-12	**In-situ finishings** (continued)	M22	**Sprayed mineral fibre coatings**	
			The rules are very similar to those for "Sand cement/.../ flooring" (M10)	
			Widths not exceeding 300mm do not have to be kept separate and are grouped in with superficial work	
			No differentiation between flat and curved work	

SMM6		SMM7	
Clause	Heading	Clause	Heading/Comment
T.4-12	<u>In-situ finishings</u> (continued)	M30	<u>Metal mesh lathing/ Anchored reinforcement for plastered coating</u>
T.4.3	Lathing, etc given separately stating type, extent of laps, method of fixing and jointing		Very little changed from the rules which would have been used in SMM6 except: a) the work has been given a specific section of its own and has its own rules
		M30:2-5. 2.*.*	b) widths not exceeding 300mm on walls, ceilings, beams and columns are measured in linear metres
		M30:7	c) Bridging to light fittings and the like are measured in linear metres
		M30:8	d) Work to irregular window and dormer cheeks are kept separate

\multicolumn{2}{c	}{SMM6}	\multicolumn{2}{c}{SMM7}
Clause	Heading	Clause	Heading/Comment
T.4-12	**In-situ finishings** (continued)	M41	**Terrazzo tiling/ in situ terrazzo**
		M41:2	In situ terrazzo is to be measured in accordance with M10

SMM6		SMM7	
Clause	Heading	Clause	Heading/Comment
T.4, 8, 9 and 11	<u>In-situ finishings</u> (continued)	Q22	<u>Coated macadam/ Asphalt roads/pavings</u>
			<u>General Rules</u>
(T.3.1)	Work classified as Internal or External under an appropriate heading	Q22:D1	Work is deemed external unless described as internal
T.4	Generally		Generally unchanged except:
		Q22:D2	Finished thickness to be stated. Widths not exceeding 300mm not kept separate. Isolated areas not kept separate
T.8	Floors	Q22:1 & 2	Roads/Pavings - refer to comments under SMM6-T.8/SMM7-M10:5 above
T.9	Channels	Q22:3	No change except arrises, coves, ends, angles, intersections and outlets are deemed to be included
T.11	Labours	Q22:C1 and Q22:C2	All deemed to be included

SMM6		SMM7	
Clause	Heading	Clause	Heading/Comment
T4, 8 and 11	<u>In-situ finishings</u> (continued)	Q26	<u>Special surfacings/ pavings for sport</u>
			This section would previously have been covered by both in-situ finishings and flexible sheet finishings or in the concrete work trade
(T.3.1)	Work classified as Internal or External under an appropriate heading	Q26:D1	Work is deemed <u>external</u> unless described as internal
T.4	Generally	Q26	Generallly unchanged - thickness is defined as the "nominal" thickness
T.8	Floors	Q26:1	Liquid applied surfacings
		Q26:4	Proprietary coloured tarmacadam sports surfacings and pavings
		Q26:5	Proprietary clay and shale coloured sports surfacings and pavings
		Q26:7	Surface dressings
			- Refer to comments under SMM6-T.8/SMM7-M10:5 above
			Widths not exceeding 300mm not differentiated but grouped in with other areas

SMM6		SMM7	
Clause	Heading	Clause	Heading/Comment
T.13	**Beds and backings**	M10	**Sand cement/Concrete/ Granolithic screeds/ flooring**
	The rules for in-situ finishings are applied (T.4-10) for beds and backings with the addition that beds and backings are to state whether screeded, floated or trowelled		In the same way as for SMM6 the rules for sand cement flooring etc. apply equally to sand cement screeds and rendering and the like

Refer to comments under SMM6 - In-situ finishings section T.4-10/SMM7-M10 above |

SMM6		SMM7	
Clause	Heading	Clause	Heading/Comment
T.14	Tile, slab or block finishings		The following work sections would previously have been measured under "Tile, slab or block finishings" or "Mosaic work"
T.15	Mosaic work		
		H51	Natural stone slab cladding/features
		H52	Cast stone slab cladding/features
		K11	Rigid sheet flooring/sheathing/ linings/casings
		M40	Stone/Concrete/ Quarry/Ceramic tiling/Mosaic
		M41	Terrazzo tiling/ In situ terrazzo
		M42	Wood block/ Composition block/ Parquet flooring
		M50	Rubber/Plastics/ Cork/Lino/Carpet tiling/sheeting (tile materials only)
		Q24	Interlocking brick/ block roads/ pavings
		Q25	Slab/Brick/Sett/ Cobble pavings
			Note: References below to H51 apply equally to H52
			Comparison of the rules for K11, M40, M41, M42, M50, Q24 and Q25 follow this section in sequence

SMM6		SMM7	
Clause	Heading	Clause	Heading/Comment
T.14	Generally		Work is deemed to include:
		H51:C1	a) fair joints b) Working around obstructions c) additional labour for overhand work d) cutting e) drainage holes f) bedding mortars and adhesives g) grouting h) cleaning, sealing and polishing
(T.3)	Generally		
(T.3.1)	Work classified as Internal or External under an appropriate heading	H51:D1	All work is deemed external unless described as internal
(T.3.2)	Work in compartments not exceeding 4.00 square metres on plan	–	No longer required to be kept separate
(T.3.3)	Overhand work	H51:C1(c)	Now deemed to be included
(T.3.4)	Grouping of work	–	No specific mention in SMM7
(T.3.5)	Patterned work so described	H51:1-13.*.*.1	To be stated in description under each classification
(T.3.6)	Nature of base	H51:S3	No change
(T.3.7)	Preparatory work	H51:S4	Now to be given in description of work
(T.3.8)	Curved work	H51:M4	Curved work shall be so described stating radii

\multicolumn{2}{c	}{SMM6}	\multicolumn{2}{c}{SMM7}
Clause	Heading	Clause	Heading/Comment
(T.3.9)	Work to areas (except staircases) over 3.50m above floors	H51:M3	No change for ceilings and beams. No requirement to give details where work to walls is over 3.50m high above floor and ceiling finish differs
(T.3.10)	Work of a repairing nature and to isolated areas not exceeding 1.00 square metres	–	No special requirements in SMM7
(T.3.11)	Temporary rules, temporary screeds and templets. Temporary support face of risers	–	No specific rules
(T.3.12)	No deductions to be made for grounds	–	No change
T.14.1	Particulars a-f	H51:S1-8	Generally as SMM6 with the addition of nature of base and any preparatory work
	g - cover fillets etc.	H51:16.3	Cover strips measured in linear metres with a dimensioned description stating method of fixing
T.14.2	Layout of joints	H51:S7	Layout of joints shall be described
T.14.3	Temporary moulds for precast tiles etc.	–	No mention of this item in SMM7
T.14.4	Work measured on the exposed face and no deduction is made for voids $\leq$ 0.50m2	H51:M1	No change

SMM6		SMM7	
Clause	Heading	Clause	Heading/Comment
T.14.5	Cutting deemed to be included	H51:C1(d)	No change
	Special tiles etc.	H51:15.1	Special units measured in linear metres as extra over with a dimensioned description and/or manufacturer's reference (Special units also include non-standard units to produce fair edges, internal and external angles, moulded and beaded edges and coved junctions.
	Corner pieces	H51:14	Corner pieces measured as above but enumerated
T.14.6	Refers to Sections T.4-12		
(T.4)	Generally		Although T.14.6 refers to T.4-12, T.4 is a general item related to in-situ finishings and T.14.1 covers general items to tile slab and block finishings
(T.5.1 -4)	Work to walls and and ceilings. Work to sides and soffits of attached beams and columns. Work to isolated beams. Work to isolated columns	H51:1-2	Walls, ceilings, isolated beams and isolated columns measured separately in square metres when exceeding 300mm wide and in linear metres when not exceeding 300mm wide A dimensioned description is to be given when work is with joints laid out to detail. Definitions of isolated or attached beams

SMM6		SMM7	
Clause	Heading	Clause	Heading/Comment
			and columns are as SMM6
(T.5.5)	External angles	H51:C2	Internal and external angles together with intersections not exceeding 10mm radius are deemed to be included
(T.5.6)	Rounded internal and external angles	H51:D3	Rounded internal and external angles exceeding 10mm radius are classified as curved work. In both the above items only non-standard units to produce internal or external angles are measurable under H51:15.1.*.0 as special units
(T.5.7)	Angle screeds etc.	H51:15	Special units - see earlier comments
(T.6)	Mouldings	H51:15	As immediately above
(T.7.1)	Staircase areas given separately	H51:M2	Work to staircase areas still measured separately
(T.7.2)	Work to landings	H51:D6	Work to floors includes landings
(T.7.3)	Work to treads and risers	H51:6-7	Treads, sills and risers to be measured linear metres stating width or height as appropriate and separating plain and undercut risers.
		H51:C4	Fair edges and internal and external angles are deemed to be included
		H51:C5	Work to curved treads and risers is deemed to

SMM6		SMM7	
Clause	Heading	Clause	Heading/Comment
			include curved and radiused cutting for special edge tiles
(T.7.4)	Nosings, fair edges etc.	H51:C4	Deemed to be included unless formed using non-standard units which are measured under H51:15.1.*.0 as special units
(T.7.5)	Strings and aprons	H51:9-10	Strings and aprons are measured separately in linear metres stating the height and are deemed to include fair edges, ends, angles and ramps
(T.8)	Floors		
(T.8.1)	Classifications		Now measured as:-
		H51:5.1	Level or to falls only not exceeding 15 degrees from horizontal
		H51:5.2	To falls and crossfalls and to slopes not exceeding 15 degrees from horizontal
		H51:5.3	To slopes exceeding 15 degrees from horizontal. All work is measured in square metres separating plain or work with joints laid out to detail and giving details of patterned work, floors laid in bays (stating average size) and inserts (stating size or sections). Work to floors is also deemed to include

SMM6		SMM7	
Clause	Heading	Clause	Heading/Comment
			intersections in sloping work
(T.8.2)	Work to boiler rooms etc.	H51:M2	Work to boiler rooms etc. is measured separately as work "in plant rooms"
(T.8.3)	Work laid in bays	H51:2.*.*.2	See comments against (T.8.1) above
(T.8.4)	Work laid in one operation with the base	H51:S6	No specific rules but covered by supplementary information
(T.9)	Channels		
(T.9.1)	Working finishings into channels	-	Not applicable to tile or slab finishings
(T.9.2)	Linings to channels	H51:11	Measured in linear metres stating girth and keeping horizontal work or work to falls separate. Deemed
		H51:C7	to include arrises, coves, ends, angles, intersections and outlets
(T.10)	Skirtings and kerbs	H51:12-13	Skirtings and kerbs are measured separately in linear metres stating the height (and width) and giving details of any patterned work, inserts (stating size or section) and whether flush, raking or
		H51:C8	vertical. Deemed to include fair edges, rounded edges, ends, angles and ramps
(T.11)	Labours		

SMM6		SMM7	
Clause	Heading	Clause	Heading/Comment
(T.11.1)	Fair joints	H51:C1(a)	Deemed to be included
(T.11.2)	Making good around pipes etc.	H51:C1(b)	Deemed to be included
(T.11.3)	Working into recessed covers, shaped inserts etc.	H51:15	Only measurable when produced using a non-standard unit
(T.12)	Dividing strips	H51:16.4	Dividing strips measured in linear metres with a dimensioned description stating method of fixing
T.14.7	Isolated special units or access units	H51:15.2 H51:15.3	Access units Isolated special units. Both enumerated with a dimensioned description and/or manufacturers reference
T.14.8	Labours as T.11		See comments earlier in this section

SMM6		SMM7	
Clause	Heading	Clause	Heading/Comment
T.14	**Tile, slab and block finishings** (continued)	K11	**Rigid sheet flooring/ sheathing/linings/ casings** As SMM6 but no requirement necessary to state layout of joints. Where work is required to isolated beams and isolated columns to be measured as girth not exceeding 600mm and thereafter in 600mm stages

SMM6		SMM7	
Clause	Heading	Clause	Heading/Comment
T.14	<u>Tile, slab and block finishings</u> (continued)	M40	<u>Stone/Concrete/ Quarry/Ceramic tiling/Mosaic</u>
		M41	<u>Terrazzo tiling/ in situ terrazzo</u>
		M42	<u>Wood block/ Composition block/ Parquet flooring</u>
			Note: References below to M40 apply equally to M41 (tiling) and M42
T.14.1 and T.14.2	Particulars to be given and layout of joints	M40:S1-8	Very similar wording. Treatment of joints is to be stated but grouting is deemed to be included (Cl(g)). The nature of the finished surface including any sealing/polishing is to be given as supplementary information but cleaning, sealing and polishing is deemed to be included. This appears to be something of an anomaly
T.14.3	Temporary moulds for precast units deemed to be included	-	Not mentioned
T.14.4	Work measured on exposed face. No deduction for voids not exceeding 0.50m2	M40:M1	No change
T.14.5	Cutting to angles, boundaries and junctions deemed to be included.	M40:Cl(d)	All cutting deemed to be included
	Corner pieces enumerated.	M40:14	No change

SMM6		SMM7	
Clause	Heading	Clause	Heading/Comment
	Special tiles, slabs or blocks given as extra over	M40:15.1-3	No change
T.14.6	Work measured in in accordance with T.4-12	M40:1 M40:2	Walls Ceilings
(T.5)	Walls, ceilings, beams and columns	M40:3	Isolated beams
		M40:4	Isolated columns
		M40:D6	Tiles now deemed to have long side vertical, unless stated horizontal
		M40:C2	All angles $\leq$ 10mm now deemed to be included
		M40:1-3.2.*.*	widths not exceeding 300mm now measured linear
T.14.6b	Work to sides and soffits of beams, columns and openings each given separately	M40:D5	Work to sides and soffits of attached beams and openings and sides of attached columns is now measured with the respective ceiling or wall
(T.6.4)	Ornaments	M40:16.5.*.*	No change
(T.7)	Staircase areas given separately	M40:M2	No change
(T.7.2)	Work to landings so described	M40:D7	Floors include landings
(T.7.3)	Treads and risers	M40:6 M40:8 M40:6.C4	Treads Risers Internal and external angles and fair edges now deemed to be included except those comprising

237

SMM6		SMM7	
Clause	Heading	Clause	Heading/Comment
		M40:15.1-3 and M40:C8	special tiles, slabs or blocks which are measured extra over
		M40:7	Sills - not mentioned as in SMM6 T.14
(T.7.4)	Nosings, fair edges, inserts and bands measured in linear metres as extra over work in which they occur	M40:C4 M40:15. 1-3 and M40:C8	Fair edges are deemed to be included except that fair edges of special tiles, slabs or blocks are measured as extra over
(T.7.5)	Strings and aprons	M40:9 M40:10 M40:C6	Strings Aprons Ends, angles and ramps deemed to be included on strings and aprons
(T.9)	Channels	M40:11.C7	Linings to channels Arrises, coves, ends, intersections and outlets now deemed to be included
(T.10)	Skirtings and kerbs	M40:12 M40:13 M40:C8	Skirtings Kerbs Fair edges, rounded edges, ends, angles and ramps now deemed to be included
(T.12)	Dividing strips	M40:16.4	Dividing strips are measured as accessories. Accessories include separating membranes, movement joints, cover strips and ornaments
T.14.7	Isolated special units and access units enumerated as extra over	M40:15.4 and M40:15.5	Access units Isolated special units No change

SMM6		SMM7	
Clause	Heading	Clause	Heading/Comment
T.14.8	Labours measured in accordance with T.11	M40:C1 M40:C1	Most labours are now deemed to be included these are: (a) fair joints (b) working over and around obstructions (c) additional labour for overhand work (d) cutting

SMM6		SMM7	
Clause	Heading	Clause	Heading/Comment
T.14	<u>Tile, slab and block finishings</u> (continued)	L42	<u>Infill panels/sheets</u>
		L42:1	To be measured in square metres with a fully detailed description
		L42:D2	Defined as non-glass and non-glass plastics rigid sheet spandrel and infill panels
		L42:C1	Glazing compounds, sealants, intumescent mastic, distance pieces, location and setting blocks and fixings are deemed to be included

SMM6		SMM7	
Clause	Heading	Clause	Heading/Comment
T.14	**Tile, slab and block finishings** (continued)	M50	**Rubber/Plastics/Cork/ Lino/Carpet tiling/ sheeting** (tile materials only) Very similar to SMM6 except: a) widths not exceeding 300mm are measured in linear metres b) work to walls over 3.50m high and where ceiling finish is dissimilar are not described separately c) external angles and rounded internal and external angles not exceeding 10mm radius are deemed to be included (C2) d) ends, angles etc. on treads, risers, skirtings etc. are deemed to be included (C5-8)

SMM6		SMM7	
Clause	Heading	Clause	Heading/Comment
T.14	<u>Tile, slab and block finishings</u> (continued)	Q24 Q25	<u>Interlocking brick/ block roads/pavings</u> <u>Slab/Brick/Block/ Sett/Cobble pavings</u>
			Note References below to Q24 apply equally to Q25
(T.3.1)	Work classified as Internal or External under an appropriate heading	Q24:D1	Work is deemed external unless otherwise described
(T.3.2)	Work in compartments not exceeding 4.00 square metres on plan	–	No longer required to be kept separate
(T.3.5)	Patterned work so described	Q24:1-2. *.*.2	No change
(T.3.6)	Nature of base	Q24:S7	No change
(T.3.7)	Preparatory work	Q24:S8	No change
(T.3.8)	Curved work	Q24:3-8. *.*.3	No change except the words conical, spherical, convex, concave, elliptical and parabolic are not used
(T.3.10)	Isolated areas not exceeding 1.00 square metres		Isolated areas not mentioned
(T.3.11)	Temporary rules, temporary screeds and templets	–	Not mentioned
T.14.1-2	Particulars to be given and layout of joints	Q24:S1-8 and D2	No change except quality of material to be given and thickness stated is defined as the "nominal" thickness
T.14.3	Temporary moulds	–	Not mentioned

SMM6		SMM7	
Clause	Heading	Clause	Heading/Comment
T.14.4	Work measured on exposed face. No deduction for voids not exceeding 0.50 square metres	Q24:M1	No change
T.14.5	Cutting deemed to be included. Special tiles measured as extra over	Q24:C1(c) Q24:9	No change No change
T.14.6	Measure in accordance with T.4-12 as appropriate except joints laid out to detail or work in one piece to be enumerated	Q24:1-10	Little change except a) widths not exceeding 300mm wide are not kept separate b) Work with joints laid out to detail to be measured in square metres with a component detail drawing (Q24:1-2.*.*.3) c) Working into shallow channels is deemed to be included (C2) d) Labours deemed to be included (C1) e) Internal and external angles are deemed to be included (C4) f) Foundation and haunching of kerbs and edgings to be included in description (S9) and formwork to same is deemed to be included (C6)

SMM6		SMM7	
Clause	Heading	Clause	Heading/Comment
T.16-19	<u>Flexible sheet finishings</u>		The following work sections would previously have been measured under "Flexible sheet finishings":
		M50	<u>Rubber/Plastics/Cork/ Lino/Carpet tiling/ sheeting</u> (sheet materials only)
		Q26	<u>Special surfacings/ pavings for sport</u> (sheet materials only)
			Comparison of the rules for Q26 follows those for M50
			General Rules:
T.16	Generally	M50:M1 and S1 to S8	No change
(T.3.1 a-b)	Work classified as Internal or External work under an appropriate heading	M50:D1	Work deemed internal unless described as external
	No requirement	M50:M2	Work in staircase areas and plant rooms kept separate
T.17 (T.5)	Wall and ceiling finishes	M50:1-4. *.*.*	No change with the exception of work which is not exceeding 300mm wide which is to be measured in linear metres
		M50:M4	Curved work measured on face
(T.3.10)	Isolated areas not exceeding 1.00 square metres kept separate	-	No requirement in SMM7 to keep isolated areas separate

SMM6		SMM7	
Clause	Heading	Clause	Heading/Comment
(T.5.6c)	Internal and external angles	M50:C2	External angles, rounded internal and external angles not exceeding 100mm radius are deemed to be included
T.18 (T.8)	Floor finishes	M50:5.*.*.*	No change except that widths not exceeding 300mm are to be measured in linear metres
	Flexible sheet finishings in SMM6 has no provision for work to staircases	M50:6-9	Work to strings, aprons, treads and risers
T.19 (T.3.3) (T.11)	Labours	M50:C1	(a) Fair joints are deemed to be included (b) Working over and around obstructions, into recesses and shaped inserts are deemed to be included
(T.10)	Skirtings and kerbs	M50:10 M50:11	Skirtings Kerbs
(T.9.2)	Channels	M50:12	Linings to channels
		M50:13	Accessories
			All these items are additional to the requirements of SMM6 for flexible sheet finishings. Had they occurred when measuring under SMM6 they would have been measured in accordance with other similar rules i.e. T.9, T.10 and T.12

SMM6		SMM7	
Clause	Heading	Clause	Heading/Comment
T.16-19	**Flexible sheet finishings** (continued)	Q26	**Special surfacings/ pavings for sport** (sheet materials only)
			Generally as SMM6 except:
		Q26:D1	a) work is deemed external unless described as internal
		Q26:D2	b) thickness is the "nominal" thickness
		Q26:C1-3	c) labours are deemed to be included
		Q26:2.1.1	d) Work not exceeding 300mm wide no longer kept separate
		Q26:8-9	Line marking/Letters and figures would previously have been measured in "Painting and decorating - SMM6 Section "V"

SMM6		SMM7	
Clause	Heading	Clause	Heading/Comment
	<u>Dry linings and partitions</u>	K10	<u>Plasterboard dry lining</u>
		K30	<u>Demountable partitions</u>
		K31	<u>Plasterboard fixed partitions/inner walls/linings</u>
			Note: References below to K10 apply equally to K31. K30 is dealt with separately following K10
T.20	Generally		
T.20.1	Location drawings to be provided except for rectangular linings without integrated services	K10:P1	Location drawings to show the scope of the work and the services within the partition or ceiling
T.20.2-3	Particulars to be given	K10:S1-11	Similar to SMM6 with the addition of the following: .S6 Thermal insulation and vapour barriers fixing with lining .S7 Insulation to limit sound transmission .S8 Moisture resistant treatment and the like .S9 Surface applications forming part of dry lining finish .S10 Isolating membranes .S11 Method of jointing composite panels

SMM6		SMM7	
Clause	Heading	Clause	Heading/Comment
		K10:M3	The above additional items are only measured with the partition where they form an integral part of the partition or are fixed thereto
T.20.4	Linings given in square metres	K10:2.1	Linings to walls to be measured in linear metres stating height in 300mm stages
		K10:2.2-3	Linings to beams and columns now mentioned separately - to be measured in linear metres giving the girth in 600mm stages and stating the number of faces
		K10:2.4	Linings to reveals and soffits of openings and recesses to be measured separately in linear metres in widths not exceeding 300mm or 300 - 600mm.
		K10:D3	Linings to reveals and soffits of openings and recesses exceeding 600mm wide are measured as walls.
		K10:2.5	Linings to ceilings to be measured in square metres
		K10:1	A new classification of "Proprietary partitions" is included to cover complete partition systems. Proprietary partitions are measured in linear metres giving thickness, height in 300mm stages and whether boarded

SMM6		SMM7	
Clause	Heading	Clause	Heading/Comment
			one or both sides. Pattern, curved and work obstructed by integral services are to be stated where applicable in all the classifications above.
			Work not exceeding 300mm wide is not kept separate
T.20.5	Measurement of sole plates, head plates, etc	K10.C3	Sole plates, head plates etc. are deemed to be included in proprietary partitions. Where not a proprietary system these items are measured in accordance with the appropriate Work Section
T.20.6	Work to isolated columns		See notes above
T.20.7	Non-rectangular linings enumerated		Not mentioned
T.20.8	Internal and external angles	K10:3-7	Angles, tee junctions, crosses and abutments measured separately,
		K10.C4-5	deemed to include extra work involved, studding, grounds, angle tape and the like
		K10:8	Fair ends measured where partition end exposed
		K10:9	Beads - measured in detail
T.20.9	Cutting to profile of openings	K10:1.4	Reveals and soffits measured separately

SMM6		SMM7	
Clause	Heading	Clause	Heading/Comment
T.20.10	Access panels	K10:11	No change
T.20.11	Labours	K10:C1	Deemed to be included
		K30	**Demountable partitions**
		K30:1	Similar to K10 proprietary partitions but actual height of partition stated and whether finish applied in the factory or on site
		K30:2	Trims (separate items fixed on site) are measured in linear metres with a dimensioned description
		K30:3	Openings for windows etc. are measured extra over the partition and include the component

SMM6		SMM7	
Clause	Heading	Clause	Heading/Comment
	Suspended ceilings, linings and support work	K40	**Suspended ceilings**
T.21.1-7	Suspended ceilings	K40:1-2	Linings to beams and ceilings to give depths of suspension $\leq$ 150mm; 150 - 500mm and thereafter in stages of 500mm otherwise very similar to SMM6
		K40:3 K40:5	Isolated strips to be measured in width $\leq$ 300mm and thereafter in 300mm stages and are defined as being strips narrower than the specified relevant lining unit dimension
		K40:6	Irregular window and dormer cheeks are enumerated and are deemed to include cutting and extra supports
		K40:7	Cavity fire barriers are measured in square metres giving the height in stages of 300mm and whether obstructed by services
T.21.5	Non-rectangular linings enumerated	-	Not mentioned
T.21.8	Trims at edges	K40:8-9	No change except now to state whether plain or floating (trims fixed to ceiling system)
		K40:11-12	Collars to service pipes and bridging are additional items

SMM6		SMM7	
Clause	Heading	Clause	Heading/Comment
T.22-28	**Fibrous plaster**	M31	**Fibrous plaster**
			General rules:
T.22.1	Particulars to be given	M31:S1-4	No change
T.22.2	Measured in square metres as carried out. No deduction for voids not exceeding 0.50 square metres. Work not exceeding 300mm so described	M31:M1	Area measured is that in contact with base. No deduction for voids not exceeding 0.50 square metres or grounds. Widths not exceeding 300mm measured in linear metres
T.23	Wall and ceiling coverings	M31:M2	Work in staircase areas and plant rooms each to be given separately
(T.3.1a -b)	Work classified as Internal or External under an appropriate heading	M31:D1	Work is deemed internal unless described as external
(T.3.2)	Work in compartments not exceeding 4.00 square metres on plan	-	No requirement
(T.3.3)	Overhand work	-	Not mentioned
(T.3.5)	Work to a pattern so described	-	No change
(T.3.6)	Nature of base	M31:S3	No change
(T.3.7)	Preparatory work	-	Not mentioned
(T.3.8)	Curved work	M31:M4	No change except various geometric shapes not named

SMM6		SMM7	
Clause	Heading	Clause	Heading/Comment
(T.3.9)	Work to ceilings and beams over 3.50m high above floor and to walls over 3.50m high where ceiling finish differs	M31:M3	Work to ceilings and beams over 3.50m high above floor unchanged. No requirement to keep walls over 3.50m high separate where ceiling finish differs
(T.3.10)	Work in repairs and isolated areas	-	Not kept separate
(T.3.11)	Temporary rules etc.	M31:C1(d)	Moulds are deemed to be included
(T.3.12)	No deduction for grounds	M31:M1	No change
T.23.1	Plain and panelled slab coverings to walls and ceilings	M31:1 M31:2	Walls Ceilings Work not exceeding 300mm wide, now measured in linear metres. Casings to beams and columns measured separately
T.23.2-5	Cutting, notches etc.	-	Not specifically mentioned (except around access panels) but assumed all will be deemed to be included
T.23.6-7	Forming openings and access panels	M31:3.1 M31:C2	Only access panels are mentioned and are measured "extra over" the work in which they occur. They are deemed to include cutting around edges and providing extra materials
T.24	Arches and domes	M31:4 M31:5 M31:6	Arches Domes Grained soffits

| \multicolumn{2}{c|}{SMM6} | \multicolumn{2}{c|}{SMM7} |

SMM6 Clause	SMM6 Heading	SMM7 Clause	SMM7 Heading/Comment
T.25	Casings to beams, columns and pilasters		
T.25.1	Plain casings	M31:7	Plain casings. No change except piers and pilasters are now classified as columns
T.25.2	Moulded and ornamental casings	M31:8	Moulded casings - No change
		M31:9	Ornamental casings - No change
T.26	Coves and cornices		
T.26.1	Coves, mouldings and cornices	M31:10	Coves
		M31:11	Mouldings
		M31:12	Cornices No change except ends, angles etc. are given as extra over (M31:M7)
T.26.2	Enrichments and Ornaments	M31:10-19. *.*.4	No change
		M31:15	No change
T.27	Consoles and canopies	M31:16	Consoles
		M31:17	Over doors
		M31:18	Canopies
		M31:19	Fireplace surrounds No change
T.28	Models	M31:20	Specially made models
		M31:21	Full size cartoons No change

SMM6		SMM7	
Clause	Heading	Clause	Heading/Comment
T.29-34	**Fitted carpeting**	M51	**Edge fixed carpeting**
T.29.1	Particulars to be given	M51:M1 and M51:S1-S8	No change
(T.3.1 a-b)	Work classified as internal or external work under an approporiate heading	M51:D1	Work is deemed internal unless described as external
(T.3.10)	Isolated areas not exceeding 1.00 square metres		No requirement to keep isolated areas seperate
T.30	Walls and ceilings (No provision for work to sides of beams and columns)	M51:1-4	Walls, ceilings and isolated beams and columns - work not exceeding 300mm wide to be measured in linear metres
T.31	Staircase areas	M51:M2	Work in staircase area and plant rooms each to be given separately
T.31.1	To landings	M51:5	It is assumed that work to landings will be included with floors
T.31.2	Carpeting to treads and risers (grouped together) in square metres stating width	M51:8 and M51:9	Treads

Risers
Treads and risers measured separately in linear metres. No mention of work to winders |
| | | M51:10 M51:11 M51:12 | Skirtings
Kerbs
Linings to channels
All these items are additional to the requirements of SMM6 for carpeting. Had they occurred when measuring under SMM6 they would have been |

SMM6		SMM7	
Clause	Heading	Clause	Heading/Comment
			measured in accordance with other similar rules (i.e. T.9 and T.10)
T.31.4	Stair rods, carpet holders, carpet clips	M51:13:6-9.*.0	No change
T.32.1	Floors	M51:5	Floors - widths not exceeding 300mm wide are to be measured in linear metres
T.32.2	Fixing at perimeter described and measured in linear metres	M51:C1(d)	Fixing at perimeter is deemed to be included
T.32.3	Cover and threshold strips	M51:13.4-5.*.0	No change
T.33	Labours	M51:C1-10	Labours generally are deemed to be included
T.34	Underlay - measured separately from floor covering	M51:5.*.*.3	Underlays are included in the description of the floor covering
	Protection		
T.35	Protecting the work	A34:1.6 A42:1.1	Dealt with in preliminaries under "Employer's requirements" and "Contractor's general cost items"

256

SMM6		SMM7	
Clause	Heading	Clause	Heading/Comment
U	**GLAZING** **Generally**	L40	**GENERAL GLAZING**
U.1	Information	L40:P1	Information is shown on location drawings under Section A Preliminaries/General Conditions
	Glass in Openings		
U.3	Generally		
U.3.1	Particulars to be given	L40:S1-5 L40:S2	No change except: Now includes cross reference to strips or channels for edges of panes L40:11
U.3.2	Panes in double glazed lights not hermetically sealed	L40:M1	Unchanged except now referred to as multiple glazed panes
U.3.3	Panes which are required to align	L40:1.*. *.6	No change
U.4	Measurement		
U.4.1	Stating the size of pane where one or both dimensions are over the manufacturers normal maximum	—	No requirement for this to be given but would suggest this should still be stated where appropriate
U.4.1.	Classification of pane sizes	L40:1.1	Pane sizes for standard plain glass now classified as follows:- 1. Panes (Nr), Area $\leq$ 0.15m2 2. Panes Area 0.15 - 4.00m2

\multicolumn{2}{SMM6}	\multicolumn{2}{SMM7}
Clause	Heading	Clause	Heading/Comment
		L40:D2	Standard plain glass is defined as any glass (other than a special glass) which is ≤ 10mm thick and in panes ≤ 4.00m2 and is not drilled; not brilliant cut and not bent
U.4.2	Panes of irregular shape	L40:1.*.*.2 L40:M3	No change
U.4.3	Where fifty or more panes are identical	L40:1.*.*.1	No change
		L40:2	Non-standard plain glass
		L40:D3	Rules are included for this classification which is defined as any glass (other than special glass) which is > 10mm thick or more or is in panes > 4.00m2 or is drilled; brilliant cut or bent
U.5	Special glass		
U.5.1.a	Glass less than 10mm thick where the size of the pane exceeds 4.00m2	L40:D3	Both now defined as non-standard plain glass
U.5.1.b	Glass 10mm thick and over		
U.5.1.c	Toughened or laminated glass	L40:3	Special glass
U.5.1.d	Solar control and other speciality glass	L40:D4	Special glasses now defined by a more comprehensive list
U.5.1.e	Acrylic, polycarbonate and similar material		
U.6	Labours on glass		
U.6.1	Raking cutting	L40:C1	No change

SMM6		SMM7	
Clause	Heading	Clause	Heading/Comment
U.6.2	Curved cutting	L40:C1	Now deemed to be included
U.6.3	Polished edges and bevelled edges	L40:5-6	No requirement now to give size of pane
U.6.4	Grinding, sandblasting and embossing	L40:7-9	No change
		L40:10	Rules now included for Engraving
U.7	Drilled panes	L40:2.*.*.8	No change
		L40:2.*.*.9	Rules now include for drilled panes with insulating sleeves
U.8	Bent panes	L40:2.*.*.5.6 & 7	No change
U.9	Brilliant cut panes	L40:2.*.*.4	No change
U.10	Hermetically sealed units	L40:3	No change
U.11	Glass louvres	L40:1.2	Standard plain glass louvres
		L40:2.2	Non-standard plain glass louvres
		L40:3.2	Special glass louvres
			No provision now for including labours to edges in description
U.12	Bedding	L40:11	No change
U.13	Hacking out	L40:13	Now required to state type of sash or other surround, method of glazing, type of glass and whether beads for re-use
		L41	**Leaded light glazing**
U.14	**Leaded lights and copper lights in openings**	L41:1	No change
		L41:2	Saddle bars to be measured

SMM6		SMM7	
Clause	Heading	Clause	Heading/Comment
		L41:2.1	Length ≤ 300mm (Nr)
		L41.2.2	Length < 300mm (Nr) No reference to copper lights - measure as for lead if need arises
U.15	Mirrors	L40:12	No change but see L40:M7 below
		L40:M7	Mirrors fixed to walls or glazed into openings are measured here. Small mirrors in toilets, dressing rooms and the like are measured in Section N10
	Patent glazing	H10	Patent glazing
		H12	Plastics glazed vaulting/walling and
		H13	Structural glass assemblies
			Note: References below to H10: apply equally to H12 and H13
U.16.1-2	Patent glazing generally	H10:1-2	Measured in square metres as SMM6 but also stating whether single or multi tiered and details of site drillings
		H10:M1	No deduction is made for voids < 1.00m2
U.16.3	Opening portions	H10:3	As SMM6 but state type of opening portion i.e. doors
U.17	Labours on patent glazing	H10:4-5	Grouped with the glazing to which they relate and measured in linear metres

SMM6		SMM7	
Clause	Heading	Clause	Heading/Comment
		H10:6	Weatherings, flashings, etc. measured in linear metres stating whether preformed or extruded, with details of site drillings to backgrounds
		H10:C2	Stop ends, mitres and corners are deemed included. Not previously covered by SMM6
	Domelights		
U.18	Generally	–	Not covered in SMM7 - If item occurs measure as appropriate, perhaps in accordance with the rules for rooflights (H14 or L11)
	Protection		
U.19	Protecting the Work		Dealt with in Preliminaries under "Employer's requirements" and "Contractor's general cost items"
		A34:1.6	
		A42:1.11	

SMM6		SMM7	
Clause	Heading	Clause	Heading/Comment
V	**PAINTING AND DECORATING**	M	**SURFACE FINISHES**
		M60	**Painting/Clear finishing**
		M52	**Decorative papers/ fabrics**
	Generally		
V.1	Information		
V.1.1	General description where not on location drawings	M60:P1 M52:P1	Information is to be shown on location drawings
V.1.2 and V.1.3	Classification of work:- Internal External Redecoration	M60:D1	Work is deemed internal unless otherwise described. No mention of redecoration work but this could be covered by the description of the base to which the decoration is to be applied
V.1.4	Work on members before fixing	M60:1.*.*.4	Included in the description of the work. Off-site priming on wood or metal before fixing is excluded from this requirement as it would be measured with the timber or metal component - refer to Code of Procedure for Measurement under M60:D2-D3
V.1.5	Work to walls ceilings and cornices in staircase areas to be stated	M60:M1 M52:M2	No change When "Painting/Clear finishing" (M60) work in plant rooms is also to be stated separately

SMM6 Clause	Heading	SMM7 Clause	Heading/Comment
V.1.6	Work to ceilings and beams over 3.50m above floor and to walls over 3.50m high where decoration of ceiling is dissimilar	M60:M4 M52:M5	Work to ceilings and beams over 3.50m above floor unchanged No requirement to keep walls over 3.50m separate
V.2	Measurement	M60:M2-3 M52:M3-4	No change
V.3-1 0	**Painting, polishing and similar work**		
V.3	Generally		
V.3.1	Particulars to be given	M60:S1-8	No change
V.3.2	Work on surfaces over 300mm girth given in square metres. Work on isolated surfaces not exceeding 300mm girth given in stages of 150mm. Work in isolated areas not exceeding 0.50m2 to be enumerated	M60:1-9.*. 1-3.*	No change except work with a girth of not exceeding 300mm is grouped together as "Isolated surfaces, girth not exceeding 300mm"
V.3.3	Mouldings and bands picked out in colours different from surrounding work to be so described	M60:D2	Such work would be described as "Multi-coloured" - refer to Code of Procedure for Measurement
V.4	General surfaces		
V.4.1	Classification	M60:D8 M60:2	General surfaces are defined as all work other than those not given specific classifications; i.e. other than: Glazed windows and screens

SMM6		SMM7	
Clause	Heading	Clause	Heading/Comment
		M60:3	Glazed sash windows
		M60:4	Glazed doors
		M60:5	Structural metalwork
		M60:6	Radiators
		M60:7	Railings, fences and gates
		M60:8	Gutters
		M60:9	Services
		M60:10	Coloured bands for coding service pipes
V.4.2	Work on friezes, coves, piers, columns and beams	-	No change
V.4.3	Corrugated, fluted and carved surfaces	M60:D4	Now grouped together as "Irregular surfaces"
V.4.4	Work in multi-colours on walls and piers or ceilings and beams	M60:D3 M60:D2	No change Refer to SMM6 V.3.3 above for extension of multi-colour definition
V.4.5	Cutting in edges on flush surfaces	M60:C2	Deemed to be included
V.5	Doors, windows, frames, linings and associated mouldings		
V.5.1	Classification of the work		
	a) Doors	M60:1 and D8	Not separately classified therefore "General surfaces"
	b) Glazed doors	M60:4	No change except panes given by area rather than 'small', 'medium', 'large' etc.
	c) Frames and linings and associated mouldings other than to windows	M60:1 and D8	Not separately classified therefore "General surfaces"

SMM6		SMM7	
Clause	Heading	Clause	Heading/Comment
	d) Windows	M60:2-3	No change except that 'sash' windows are kept separate
	e) Work on edges of opening casements	M60:C4	Now deemed to be included
	f) Work on glazing beads, butts and fasteners deemed to be included	M60:C4	No change
V.6	Structural metalwork	M60:5	Structural metalwork now classified as "general surfaces" or "members of trusses, lattice girders etc". Where over 5.0m above floor level, now separately classified at 3.0m intervals of height measured to the highest point of the member
		M60:M8	
V.7	Radiators	M60:6	Radiators. Now classified as "panel" or "column" type
V.8	Railings, fences and gates	M60:7	No change. Measurement rules M10-12 explains what to measure for the particular types. Definition rules D10-11 define plain open type and close type fencing
V.9	Gutters	M60:8	No change
V.10	Pipes, ducts etc.	M60:9	Services - not now split into the various types of pipe, conduit, etc., otherwise unchanged

| \multicolumn{2}{c|}{SMM6} | \multicolumn{2}{c}{SMM7} |

Clause	Heading	Clause	Heading/Comment
	Signwriting	N15	**Signs/Notices**
V.11	Generally	N15:2.1	Signwriting
			Supplementary Information S1 not so specific as SMM6. Details as required by SMM6 V.11 would be required or such other information as is appropriate to properly describe the work
		Q26:8	Line Marking Widths less than 300mm are grouped together, widths over 300mm are measured separately stating the width
		Q26:9	Letters and figures - no change
V.12	**Decorative paper, sheet plastic or fabric backing and lining**	M52	**Decorative papers/ fabrics**
			Generally unchanged except as follows:-
V.12.1	Generally	M52:S1-4	Width and length of roll and type of pattern now to be stated
V.12.2	Walls and columns; Ceilings and beams	M52:1-2.1.0.1	Raking and curved cutting to be stated in description
		M52:1-2.2.0.*	Areas not exceeding 0.50m2 to be enumerated

SMM6		SMM7	
Clause	Heading	Clause	Heading/Comment
	Protection		
V.13	Protecting the work		Dealt with in preliminaries under "Employer's requirements" and "Contractor's general cost items"
		A34:1-6	
		A42:1.11	

SMM6		SMM7	
Clause	Heading	Clause	Heading/Comment
W	**DRAINAGE**	R	**DISPOSAL SYSTEMS**
		R12	**Drainage Below Ground**
		R13	**Land Drainage**
			Note: References below to R12 apply equally to R13
	Generally		
W.1	Information	R12:P1-2	Information reworded with more details in measurement rules, coverage rules and supplementary information. See also General Rules Clause 4.5
W.2	Plant	A43	Dealt with in Preliminaries
	Pipe trenches		
W.3	Excavation		
W.3.1	Trench starting level and depths	R12:1.*.*.1	Commencing level stated only when greater than 0.25m below existing ground level, average depth classification unchanged. Depth range not required to be given
W.3.2	Curved trenches so described	R12:1.*.*.2	No change
W.3.3	Size classification	R12:1.1	No change
W.3.4	Earthwork support treating bottoms filling and disposal	R12:C1 R12:D3	Now deemed to be included with trenches except for backfilling with special materials,

SMM6		SMM7	
Clause	Heading	Clause	Heading/Comment
			specific disposal and multiple handling and surface treatments which are given in description of trench in accordance with Work Section D20
W.3.5	Excavation below ground water	R12:M3	Now measured as excavating trenches below ground water level - not extra over
W.3.6	Trenches next roadway, existing buildings or unstable ground	R12:D2	No change except no requirement to measure excavation in unstable ground for full depth but we consider it should still apply
W.3.7	Trenches in rock	R12:2.1.1	Now only given in cubic metres as extra over excavation. Measurement shall be subject to a minimum width of 500mm
W.3.8	Breaking up concrete etc.	R12:2.1-2.2-5	Now only given in cubic or square metres as extra over excavation. Measurement shall be subject to a minimum width of 500mm
	Additional items to trenches etc.	R12:2.4-5.1	Excavating next to or across existing live services where specifically required is to be given as extra over the trench excavation, stating the type of service

SMM6		SMM7	
Clause	Heading	Clause	Heading/Comment
W.4	Disposal of water	R12:M6	Only measured if trench excavation below water level is also measured
W.5	Beds, benchings and coverings	R12:4-7	Haunchings and surrounds now given as combined items with beds. Formwork now deemed included
W.6	Pipework		
W.6.1	Pipes	R12:8	No change
W.6.2	Suspended pipes	R12:8.4	Pipe supports are now deemed to be included. Pipes over 3.50m above floor level are kept separate
W.6.3	Pipes in runs not exceeding 3m	R12:8.*.*.1	Now only applicable to iron pipes
W.6.4	Vertical pipes	R12:8	No change
W.6.5	Pipe fittings	R12:9 & S5	No change except method of jointing to pipe now required to be stated
W.6.6	Accessories	R12:10	No change except jointing to pipe and bedding in concrete now deemed to be included
	Manholes, soakaways, cesspits and septic tanks		
W.7	Generally		
W.7.1	Inspection chambers, manholes, soakaways and the like	R12:11-13	Assume to be given under an appropriate heading stating the number. No provision for grouping together manholes, inspection

SMM6		SMM7	
Clause	Heading	Clause	Heading/Comment
-	Additional items to manholes etc.	R12:11-16. 14	chambers, etc. Preformed manholes inspection chambers, soakaways, cesspits and septic tanks are to be be enumerated giving full details
W.7.2	Cesspits, septic tanks and the like	R12:14-15	As immediately above
W.7.3	Excavation, concrete, brickwork and other work not covered	R12:M8	No change
W.7.4	Building in ends	R12:11-15. 7 & C6	No change except only measured separately in non-preformed systems
W.7.5	Channels, benchings, step irons, covers and intercepting traps	R12:11-15. 7-13 & M9	Now only measured separately in non-preformed systems
W.7.6	Work to existing manholes etc.	-	No specific coverage. Suggest measure in accordance with the rules for breaking into existing pipes - refer to Additional rules - R10-13 Drainage - work to existing buildings
	Connections to Sewers		
W.8	Generally	R12:16 & M10	To be measured in this section only if carried out by the Contractor. If to be executed by a Statutory Authority then is to be given as a Provisional Sum

SMM6		SMM7	
Clause	Heading	Clause	Heading/Comment
			in accordance with A53 - Preliminaries and general conditions
	Testing drains		
W.9	Generally	R12:17	Testing and commissioning. Separate items to be given for each installation stating details of preparatory operations, stage tests, Insurance company tests and Instructions to operating personnel. Provision of water and other supplies and provision of test certificates is deemed included
-	Additional items generally	R12:18	Preparing drawings giving information as stated
		R12:19	Operating and maintenance manuals giving information as stated
	Protection		
W.10	Protecting the work		Dealt with in Preliminaries under
		A34:1.6	"Employer's requirements" and
		A42:1.11	"Contractor's general cost items"

272

SMM6		SMM7	
Clause	Heading	Clause	Heading/Comment
		Q	**PAVING/PLANTING/ FENCING/SITE FURNITURE**
X	**FENCING**	Q40	**FENCING**
	Generally		
X.1	Information		
X.1.1	General description where not on location drawings	Q40:P1	Work to be shown on location drawings
X.1.2.a	Kind and quality of material	Q40:S1-4	Kind and quality of materials, construction, surface treatments applied before delivery, size and nature of backfilling
X.1.2.b	Preliminary treatment of material		
X.2-7	**Open type fencing**	)Q40:1)	Measurement rules no longer divided into the various types of fencing. The type is to be stated in the description together with the height of fencing, the height, spacing and depth of supports. Curved fencing is to be classified by radius not exceeding or exceeding 100m. Fencing to ground sloping exceeding 15 degrees from the horizontal is to be so described. Short lengths not exceeding 3m to be separately described
X.8-10	**Close type fencing**	)	

SMM6		SMM7	
Clause	Heading	Clause	Heading/Comment
	Gates		
X.11.1	Gates		Gates. No change except:-
		Q40:5	Ironmongery is to be
		Q40:6	enumerated separately in accordance with Section P21 stating type, kind, quality, finish, method of fixing, nature of background.
	Gate posts (other than part of fencing)	Q40:3	Independent gate posts; height, depth and type required
X.11.2	Gate stops etc.	Q40:C5	Gate stops, catches, stays and associated works are deemed to be included
	Sundries		
X.12	Concrete spurs	Q40:2	Special supports extra over fencing in which they occur
X.13	End-posts, angle posts, gate-posts, straining-posts and other posts		.1 End posts; .2 Angle posts; .3 Integral gate; posts; .4 Straining posts; .5 Other details stated No change except holes for posts are deemed to be included
X.14	Holes and mortices for posts		-
X.14.1	Holes for posts	Q40:C1	Deemed to be included
X.14.2	Mortices for posts etc. in concrete, brickwork and the like	Q40:2.*. *.1	Method of fixing to background and the type of background is to be stated

SMM6		SMM7	
Clause	Heading	Clause	Heading/Comment
		Q40:4	Extra over items have been included for:- .1 Excavating below ground water level (m3) .2 Breaking out existing materials (m3) .3 Breaking out existing hard pavings (m2)
	Protection		
X.15	Protecting the work	A34:1.6 A42:1.11	Dealt with in preliminaries under "Employer's requirements" and "Contractor's general cost items"

4 : CESMM2
– How it Compares with SMM7

The second edition of the Civil Engineering Standard Method of Measurement (CESMM2) was first published by the Institution of Civil Engineers in 1985.

It was the first method of measurement to use the tabulated layout dividing the work into Work Classifications with three divisions of description and measurement, definition, coverage and additional description rules.

The seventh edition of the Standard Method of Measurement of Building Works (SMM7) has been set out in a tabulated layout very similar to that used for CESMM2.

The rules to be used for construction activities which appear in both documents (e.g. Piling) are however quite different. To assist those measuring in accordance with both methods of measurement we have prepared a comparison between the two documents. It was not our intention to compile this comparison in the same detail as that prepared to compare SMM7 with SMM6 but we have endeavoured to highlight the major differences.

There are comparable methods of measurement for some Work Sections of SMM7 and the classes of CESMM2 and these are as set out below:-

SMM7 WORK SECTION	CESMM2 CLASS
A Preliminaries/General Conditions	A General Items
C Demolition/Alteration/ Renovation	D Demolition and Site Clearance
C10 Demolishing structures	
D Ground Work	
D10 Ground investigation	B Ground investigation
D11 Soil stabilisation	C Geotechnical and other specialist processes (C1-5 and 7 and 8)
D12 Site dewatering	
D20 Excavation/Filling	E Earthworks (E1 - 7)
D30 Cast in place concrete piling	P Piles
D31 Preformed concrete piling	Q Piling ancillaries
D32 Steel piling	
D40 Diaphragm walling	C Geotechnical and other specialist processes (C6)
E In situ concrete/Large precast concrete	
E10 In situ concrete	F Insitu concrete (F1-6)
E11 Gun applied concrete	
E20 Formwork for in situ concrete	G Concrete ancillaries (G1-4)
E30 Reinforcement for in situ concrete	)G Concrete ancillaries) (G5 & 7)
E31 Post tensioned reinforcement for in situ concrete	))))

SMM7 WORK SECTION			CESMM2 CLASS
	E40	Designed joints in situ concrete	)G Concrete ancillaries) (G6 & 8))
	E41	Worked finishes/Cutting to in situ concrete	)))
	E42	Accessories cast into in situ concrete	))
	E50	Precast concrete large units	H Precast concrete (H1-8)
F	Masonry		U Brickwork, blockwork and masonry Note:- For precast concrete sills/lintels/ copings/features see Class H
G	Structural/Carcassing metal/timber		
	G10	Structural steel framing	)M Structural metalwork) (M1-8))
	G11	Structural aluminium framing	))
	G12	Isolated structural metal members	N Miscellaneous metalwork (N1 & 2)
	G20	Carpentry/Timber framing/First fixing	O Timber (O1, 2 & 5)
	G30	Metal profiled sheet decking	N Miscellaneous metalwork (N1 & 2)
	G31	Prefabricated timber unit decking	O Timber (O3 & 4)
J	Waterproofing		))
	J20	Mastic asphalt tanking/damp proof membranes	)W Waterproofing))
	J21	Mastic asphalt roofing/insulation/finishes	)))
	J30	Liquid applied tanking/damp proof membranes	))

SMM7 WORK SECTION		CESMM2 CLASS	
J31	Liquid applied waterproof roof coatings	W	Waterproofing
J32	Sprayed vapour barriers		
J33	In situ glass reinforced plastics		
J40	Flexible sheet tanking/ damp proof membranes		
J41	Built up felt roof coverings		
J42	Single layer plastics roof coverings		
M	Surface finishes		
M60	Painting/Clear finishing	V	Painting
Q	Paving/Planting/Fencing/Site furniture		
Q10	Stone/Concrete/Brick/ kerbs/edgings/channels		
Q20	Hardcore/Granular/ Cement bound bases/ sub-bases to roads/ pavings		
Q21	In situ concrete roads/ pavings/bases	R	Roads and pavings
Q22	Coated macadam/ asphalt roads/pavings		
Q23	Gravel/Hoggin roads/ pavings		
Q24	Interlocking brick/ block roads/pavings		
Q25	Slab/Brick/Block/Sett/ Cobble pavings		
Q30	Seeding/Turfing	E	Earthworks (E8)
Q31	Planting		

SMM7 WORK SECTION	CESMM2 CLASS
Q40 Fencing	X Miscellaneous work (X1 & 2)
R Disposal Systems	I Pipework - pipes
	J Pipework - fittings and valves
	K Pipework - manholes and pipe ancillaries
	L Pipework - supports and protection, ancillaries to laying and excavation
	X Miscellaneous work (X3)

Having covered the SMM7 Work Sections and CESMM2 classes which are comparable we are left with a small number of CESMM2 classes for which there are no equivalent SMM7 Work Sections and a somewhat larger number of SMM7 Work Sections for which there are no equivalent CESMM2 classes.

The CESMM2 classes for which there are no comparable work sections are as set out below:-

CESMM2
CLASS

S   Rail track

T   Tunnels

X   Miscellaneous work (X4 - rock filled gabions)

Y   Sewer renovation and ancillary works

The larger number of SMM7 Work Sections for which there are no equivalent CESMM2 classes are as set out below:-

SMM7 WORK
GROUP/SECTION

B	Complete buildings
C	Demolition/Alteration/Renovation
	C20 Alterations - spot items
	C30 Shoring
	C40 Repairing/Renovating concrete/brick/block/stone
	C41 Chemical dpcs to existing walls
	C50 Repairing/Renovating metal
	C51 Repairing/Renovating timber
	C52 Fungus/Beetle eradication
D	Groundwork
	D50 Underpinning
E	In situ concrete/Large precast concrete
	E60 Precast/Composite concrete decking
G	Structural/Carcassing metal/timber
	G32 Edge supported/Reinforced woodwool slab decking
H	Cladding/Covering
J	Waterproofing
	J10 Specialist waterproof rendering
	J22 Proprietary roof decking with asphalt finish
	J43 Proprietary roof decking with felt finish
K	Linings/Sheathing/Dry Partitioning
L	Windows/Doors/Stairs

281

SMM7 WORK
GROUP/SECTION (continued)

- M   Surface finishes

    - M10   Sand cement/Concrete/Granolithic screeds/flooring
    - M11   Mastic asphalt flooring
    - M12   Trowelled bitumen/resin/rubber-latex flooring
    - M20   Plastered/Rendered/Roughcast coatings
    - M21   Insulation with rendered finish
    - M22   Sprayed mineral fibre coatings
    - M23   Resin bound mineral coatings
    - M30   Metal mesh latching/Anchored reinforcement for plastered coatings
    - M31   Fibrous plaster
    - M40   Stone/Concrete/Quarry/Ceramic tiling/Mosaic
    - M41   Terrazzo tiling/in situ terrazzo - tiling only
    -   - in situ only
    - M42   Wood block/Composition block/Parquet flooring
    - M50   Rubber/Plastics/Cork/Lino/Carpet tiling/sheeting
    - M51   Edge fixed carpeting
    - M52   Decorative papers/fabrics

- N   Furniture/Equipment
- P   Building fabric sundries
- Q   Paving/Planting/Fencing/site furniture

    - Q26   Special surfacings/pavings for sport
    - Q50   Site/Street furniture/equipment

- X   Transport systems
- Y   Services reference specification/measurement
- -   Additional rules - work to existing buildings

Having given a very brief comparison of SMM7 Work Sections and CESMM2 classes the remaining pages of this chapter set out the principal comparable items in more detail. Some of the sub-divisions of comparable Work Sections have no similar items in

CESMM2, in this instance the sub-division wording has been printed and the CESMM2 columns have been left blank. Where the wording of an item in CESMM2 is the same as that in SMM7 the SMM7 columns have been completed and only the CESMM2 clause reference has been stated.

The comparison which follows has been set out in SMM7 Work Section order. To assist in finding particular CESMM2 clause references a list of CESMM2 classes in alphabetical order is included as Appendix B at the end of this book.

SMM7		CESMM2	
Clause	Heading	Clause	Heading/Comment
A	**PRELIMINARIES/ GENERAL CONDITIONS**	A	**GENERAL ITEMS** Note:- Many of the SMM7 sub-sections opposite have no comparable items in this Class but, they will be dealt with when preparing the Contract and Specification
A10	Project particulars		
A11	Drawings		
A12	The site/Existing buildings		
A13	Description of the work		
A20	Form of Contract	A 1	Contractual requirements
A20:1.1	Schedule of clause headings of standard conditions	A 1 2	Insurance of the Works
		A 1 3	Insurance of constructional plant
		A 1 4	Insurance against damage to persons and property
A20:1.2	Details of special conditions or amendments to standard conditions		
A20:1.3	Details of appendix insertions		
A20:1.4	Employer's insurance responsibility		

SMM7		CESMM2	
Clause	Heading	Clause	Heading/Comment
A20:1.5	Performance guarantee bond	A 1 1	Performance bond
A30	Employer's requirements; Tendering/ Sub-letting/supply		
A31	Employer's requirements; Provision, content and use of documents		
A32	Employer's requirements; Management of the Works		
A33	Employer's requirements; Quality Standards/ Control		
A34	Employer's requirements; Security/Safety/ Protection		
A34:1.1	Noise and pollution control		
A34:1.2	Maintain adjoining buildings		
A34:1.3	Maintain public and private roads		
A34:1.4	Maintain live services		
A34:1.5	Security		
A34:1.6	Protection of work in all sections		
A34:1.7	Others		

SMM7		CESMM2	
Clause	Heading	Clause	Heading/Comment
A35	Employer's requirements; specific limitations on method/sequence/timing		
A35:1.1	Design constraint		
A35:1.2	Method and sequence of work		
A35:1.3	Access		
A35:1.4	Use of the site		
A35:1.5	Use or disposal of materials found		
A35:1.6	Start of work		
A35:1.7	Working hours		
A35:1.8	Others		
A36	Employer's requirements; Facilities/Temporary works/services; Details stated		
A36:1.1	Offices	A 2 1	Accommodation for Engineer's staff
		A 3 1	Acommodation and buildings
A36:1.2	Sanitary accommodation		
A36:1.3	Temporary fences, hoardings, screens and roofs	A 3 2 4	Hoardings
A36:1.4	Name boards		
A36:1.5	Technical and surveying equipment		

SMM7		CESMM2	
Clause	Heading	Clause	Heading/Comment
A36:1:6	Temperature and humidity		
A36:1.7	Telephone/ Facsimile installations and rental/ maintenance		
A36:1.8	Others	A 2 2	Services for the Engineer's staff
		A 2 3	Equipment for use by the Engineer's staff
		A 2 4	Attendance upon Engineer's staff
		A 2 5	Testing of materials
		A 2 6	Testing of the works
		A 2 7	Temporary Works
A36:1.9	Telephone/ Facsimile call charges		
A37	Employer's Requirements; Operation/ Maintenance of the finished building		
A40	Contractor's general cost items; Management and Staff	A 3 7	Supervision and labour
A41	Contractor's general cost items; Site accommodation	A 3 1	Accommodation and buildings
A42	Contractor's general cost items; Services and facilities		
A42:1.1	Power	A 3 2 1	Electricity
A42:1.2	Lighting	A 3 2 1	Electricity
A42:1.3	Fuels		

SMM7		CESMM2	
Clause	Heading	Clause	Heading/Comment
A42:1.4	Water	A 3 2 2	
A42:1.5	Telephone and administration		
A42:1.6	Safety, health and welfare	A 3 2 7	Welfare
A42:1.7	Storage of materials		
A42:1.8	Rubbish disposal		
A42:1.9	Cleaning		
A42:1.10	Drying out		
A42:1.11	Protection of work in all sections		
A41:1.12	Security	A 3 2 3	Security
A32:1.13	Maintain public and private roads		
A42:1.14	Small plant and tools		
A42:1.15	Others, details stated		
A42:1.16	General attendance on nominated sub-contractors		
A43	Contractor's general cost items; Mechanical plant		
A43:1.1	Cranes	A 3 3 1	Cranes
A43:1.2	Hoists		
A43:1.3	Personnel transport	A 3 2 6	Personnel transport
A43:1.4	Transport	A 3 2 5 A 3 3 2	Site transport Transport

SMM7		CESMM2	
Clause	Heading	Clause	Heading/Comment
A43:1.5	Earthmoving plant	A 3 3 3 A 3 3 4	Earthmoving Compaction
A43:1.6	Concrete plant	A 3 3 5 A 3 3 6	Concrete mixing Concrete transport
A43:1.7	Piling plant	A 3 3 7 A 3 3 8	Pile driving Pile boring
A43:1.8	Paving and surfacing plant	A 3 4 2	Paving plant
A43:1.9	Others, details stated	A 3 4 1	Pipelaying plant
		A 3 4 3	Tunnelling plant
		A 3 4 4	Crushing and screening plant
		A 3 4 5	Boring and drilling plant
A44	Contractor's general cost items; Temporary works		
A44:1.1	Temporary roads	A 3 5 3	Access roads
A44:1.2	Temporary walkways		
A44:1.3	Access scaffolding	A 3 6 1	Access scaffolding
A44:1.4	Support scaffolding and propping	A 3 6 2	Support scaffolding and propping
A44:1.5	Hoardings, fans, fencing etc.		
A44:1.6	Hardstanding	A 3 6 6	Hardstandings
A44:1.7	Traffic Regulations	A 3 5 2	Traffic regulation
A44:1.8	Others, details stated		
A50	Work/Materials by the Employer		

SMM7		CESMM2	
Clause	Heading	Clause	Heading/Comment
A50:1.1	Work by others directly employed by the Employer, details stated		
A51	Nominated Sub-contractors	A 5	Nominated Sub-contracts which include work on the site
		A 6	Nominated sub-contracts which do include work on the site
A51:1.1	Sub-contractor's work	A 5 or 6.1	Prime cost item
A51:1.2	Main contractor's profit	A 5 or 6.4	Other charges and profit
A51:1.3	Special attedance, details stated	A 5 or 6.2	Labours
		A 5 or 6.3	Special labours
A52	Nominated Suppliers		
A52:1.1	Supplier's materials		
A52:1.2	Main contractor's profit		
A53	Work by statutory authorities		
A53:1.1	Work by Local Authority		
A53:1.2	Work by statutory undertakings		
A54	Provisional work	A 4 2	Other provisional sums
A54:1.1	Defined		
A54:1.2	Undefined		
A55	Dayworks	A 4 1	Daywork
A55:1.1	Labour		

SMM7		CESMM2	
Clause	Heading	Clause	Heading/Comment
A55:1.2	Materials		
A55:1.3	Plant		
	Comment		
	Both methods of measurement provide for fixed and time related charges.		
	CESMM2 also makes provision for method related charges whereby a tenderer may insert sums to cover items of work relating to his intended method of executing the works, the costs of which are not considered as proportional to the quantities of other items for which he has not allowed in the rates and prices of the other items.		
C	**DEMOLITION/ ALTERATION/ RENOVATION**	D	**DEMOLITIONS AND SITE CLEARANCE**
C10:1	Demolishing all structures	D 1	General clearance in hectares
C10:2	Demolishing individual structures	D 4 & 5	Predominant materials to be stated and volume of building to be given
		D 6	Rules for demolition of pipe lines
			There are no comparable items for the remaining classifications in Section C of SMM7

SMM7		CESMM2	
Clause	Heading	Clause	Heading/Comment
D	**GROUNDWORK**		
D10	Ground investigation	B	**GROUND INVESTIGATION**
		C	**GEOTECHNICAL AND OTHER SPECIALIST PROCESSES**
D11	Soil stabilisation	C 1-5	Drilling and grouting
D12	Site dewatering	C 7 & 8	Ground anchorages, sand band and wick drains
	No rules prepared for D10-12		Comprehensive rules included for B, C1-5 and C7 & 8
			DEMOLITION AND SITE CLEARANCE
D20:1.1	Removing trees	D 2	Removing trees girth 500mm - 1m 1 - 2m 2 - 3m 3 - 5m exceeding 5m
D20:1.2	Removing tree stumps	D 3	Removing stumps diameter 150 - 500mm 500mm - 1m exceeding 1m
		E	**EARTHWORKS**
D20:2.1	Top soil for preservation average depth to be given	E 4 1	General excavation - top soil maximum depth classifications as below apply

SMM7		CESMM2	
Clause	Heading	Clause	Heading/Comment
D20:2.2 to D20:2.7	Reduce levels Basements Pits Trenches Pile caps and ground beams between piles		Reduce levels classified as general excavation remainder classified as excavation for foundations by maximum depth: not exceeding 0.25m 0.25 - 0.5m 0.5 - 1m 1 - 2m 2 - 5m 5 - 10m 10 - 15m stated exceeding 15m
D20:4&5	Breaking out existing materials		Excavation in various soils and materials kept separate as follows:- 1. Top soil 2. Material other than top soil rock or artificial hard material 3. Rock 4. Stated artificial hard material exposed at the commencing surface 5. Stated artificial hard material not exposed at the commencing surface
D20:6	Working space allowance to excavations		Deemed to be included
D20:7	Earthwork support	E 5 7 & 8	Timber and metal supports only measured when left in

SMM7		CESMM2	
Clause	Heading	Clause	Heading/Comment

D20:9	Filling to excavations	E 6	Filling to:-
			  structures m3
			  embankments m3

D20:10	Filling to make up to levels		general m3
			  to stated depth
			  thickness m2

			and classified as follows:-
			1. Excavated top soil 2. Imported top soil 3. Non-selected excavated material other than top soil or rock 4. Selected excavated material other than top soil or rock 5. Imported natural material other than top soil or rock 6. Excavated rock 7. Imported rock 8. Imported artificial material
D20:13	Surface treatments	E 5 E 7	Excavation ancillaries Filling ancillaries Blinding surface of filling not measurable Trimming and preparation are separately measured for the various soils and materials
D30	**Cast in place piling**	P Q	**PILES** **PILING ANCILLARIES**
D30:1 & 2	Bored piles and driven shell piles	P 1 & 2	
D30:3	Pre-boring driven piles	Q 3 1	Cross-sectional area to be given in lieu of nominal diameter

SMM7		CESMM2	
Clause	Heading	Clause	Heading/Comment
D30:4	Backfilling empty bores	Q 1 2	
D30:5.1	Breaking through obstructions	Q 7	
D30:5.2 & 3	Enlarged bases	Q 1 5	
D30:6	Permanenet casings	Q 1 3	Lengths not exceeding 13m and exceeding 13m to be kept separate
D30:7	Cutting off tops of piles	Q 1 7	Cutting off surplus lengths
	deemed to include preparation	Q 1 8	Preparing heads
D30:8	Reinforcement to piles	Q 2 1	Sizes of bars are grouped together as not exceeding 25mm and exceeding 25mm. No provision for reinforcement to be included with piles up to 600mm diameter as SMM7
D30:9	Disposal	Q C 1	Disposal of surplus materials deemed to be included
D30:10	Delays		
D30:11	Pile Tests	Q 8	More specific rules provided
D31	**Preformed concrete piling**		
D31:1	Reinforced piles	P 3	Preformed concrete piles
D31:2	Prestressed piles	P 4	Preformed prestressed concrete piles
D31:3	Reinforced sheet piles	P 5	Preformed concrete sheet piles

SMM7		CESMM2	
Clause	Heading	Clause	Heading/Comment
D31:4	Hollow section piles		
D31:5	Items extra over piling		
D31:6	Pre-boring	Q 3 1	
D31:7	Jetting	Q 3 2	
D31:8	Filling hollow piles with concrete	Q 3 3	
D31:9	Pile extensions	Q 3 4 Q 3 5 Q 3 6	Lengths not exceeding 3m and exceeding 3m to to be kept separate
		P A 8	Details of shoes shall be stated in item descriptions for the number of piles
D31:10	Cutting off tops of piles	Q 3 7	Cutting of surplus lengths
	deemed to include preparation	Q 3 8	Preparing heads
D31:11	Disposal	Q C 1	Disposal of surplus material deemed to be included
D31.12	Delays		
D31:13	Piles tests	Q 8	More specific rules provided
D32	**Steel piling**		
D32:1	Isolated piles	P 7	Isolated steel piles
D32:2	Interlocking piles	P 8	Interlocking steel piles

D32:3	Items extra over interlocking piles
	  corners
	  junctions
	  closures
	  tapers
		P D 7	Corner junction closure and tapers to be closed as special piles

SMM7		CESMM2	
Clause	Heading	Clause	Heading/Comment
D32:4	Isolated pile extensions	Q 5 4-6 5 & 6	
D32:5	Interlocking pile extensions	Q 6 4-6 5 & 6	
D32:6	Cutting off surplus from specified lengths		
D32:6	Isolated piles	Q 5 7	Cutting off surplus lengths
D32:6.2	Interlocking piles	Q 6 7	Cutting off surplus lengths
D32:7	Cutting interlocking piles to form holes		
D32:8	Delays		
D32:9	Piles tests	Q 8	More specific rules provided
		P 6	Timber piles
D40	**Diaphragm Walling**	C	**GEOTECHNICAL AND OTHER SPECIALIST PROCESSES**
D40:1	Excavation and disposal	C 6 1-3	Maximum depth in stages
D40:2	Items extra over excavation		Excavation in various soils and materials kept separate
			1. Excavation in material other than rock or artifical hard material
			2. Excavation in rock
			3. Excavation in artificial hard material

SMM7		CESMM2	
Clause	Heading	Clause	Heading/Comment
D40:3	Backfilling empty trench		
D40:4	Concrete	C 6 4	
D40:5	Reinforcement	C 6 5 & 6	
D40:6	Cutting off to specified level		deemed to be included - see Coverage Rule C3
D40:7	Trimming and clearing face of diaphragm wall		deemed to be included - see Coverage Rule C3
D40:8	Waterproofed joints	C 6 7	
D40:9	Guide walls one side both sides	C 6 8	guide walls shall be measured each side - see Measurement Rules M12
D40:10	Ancillary work in connection with diaphragm walling		No specific provision made for these items
D40:11	Delays		No specific provision made for these items
D40:12	Tests		Testing of the Works

SMM7		CESMM2	
Clause	Heading	Clause	Heading/Comment
E	IN SITU CONCRETE/ LARGE PRECAST CONCRETE		
E10	In situ Concrete	F	IN SITU CONCRETE
E10:1	Foundations	F 5 2	Bases footings pile caps
E10:2	Ground beams		Thickness to be stated classified as follows:-

E10:3	Isolated foundations		not exceeding 150mm
			  150 - 300mm
			  300 - 500mm
			  exceeding 500mm

E10:4	Beds	F 5 2	Ground slabs) Thickness) to be
E10:5	Slabs	F 5 3	Suspended) stated slabs) classified
E10:6	Coffered and troughed slabs		) as above) which) varies
E10:7	Walls	F 5 4	Walls) from SMM7
E10:8	Filling hollow walls		
E10:9	Beams	F 5 6	Beams) Cross-) sectional
E10:10	Beam casings	F 5 7	Casing to) areas of metal) these sections) items to
E10:11	Columns	F 5 5	Columns) be given and) - not piers) required
E10:12	Column casings	F 5 7	Casing to) by SMM7 metal) sections)
			The provision of concrete is to be given as separate items only the placing of concrete is to be given as above
E10:13 to E10:17			No equal provision made for these items in CESMM2

SMM7		CESMM2	
Clause	Heading	Clause	Heading/Comment
E20	**Formwork for in-situ concrete**	G 1-4	**CONCRETE ANCILLARIES** Note: Classifications of position within the works i.e. foundations, edges of beds etc. not included
	Classification of surfaces:-		
E20:1.1	Plain vertical	G 1 1	Plane vertical
E20:8.1 .1	Horizontal	G 1 4	Plane horizontal
E20:8.1 .2	Sloping not exceeding 15 degrees	G 1 2	Plane sloping
E20:8.1 .3	Sloping exceeding 15 degrees		
		G 1 3	Plane battering
M2	Curved work to be so described with the radius stated	G 1 5	Curved to one radius in one plane
		G 1 6	Other curved
	Note:- both methods provide for formwork to be measured to sloping upper surfaces exceeding 15 degrees and where specifically required		
	Classification of heights given as widths in CESMM2:-		
E20:1.1 .1	Height > 1.00m - m2		
E20:1.1 .2	Height ≤ 250mm - m	G 1 1 1	Width; not exceeding 0.1m -m
E20:1.1 .3	Height 250 - 500mm -m	G 1 1 2	0.1 - 0.2m -m
E20:1.1 .4	Height 500 - 1.00m -m	G 1 1 3	0.2 - 0.4m -m2
		G 1 1 4	0.2 - 1.22m -m2
		G 1 1 5	exceeding 1.22m -m2
	Propping heights:-		No provision made for stating heights of propping
E20:8.1 .1.1	≤ 1.50		
E20:8.1 .1.2	thereafter in 1.50m stages		

SMM7		CESMM2	
Clause	Heading	Clause	Heading/Comment
	Slab thicknesses to be stated for formwork to soffits of slabs:-		No provision included for stating thickness of slabs
E20:8.1	$\leq$ 200mm		
E20:8.2	and thereafter in 100mm stages		
		G 1 8	Concrete components of constant cross-section:-
E20:12	Walls	G 1 8 3	Walls
E20:13	Beams	G 1 8 1	Beams
E20:14	Beams casings	G 1 8 4	Other members
E20:15	Columns	G 1 8 2	Columns
E20:16	Column Casings	G 1 8 4	Other members
E20:17	Recesses	G 1 8 6	Intrusions
E20:18	Nibs	G 1 8 5	Projections
E20:19	Rebates	G 1 8 6	Intrusions
E20:20	Extra over a basic finish for a formed finish	G 1	Formwork: rough finish fair finish other stated finish stated surface features
E26	Mortices	G 1 7	For voids
E27	Holes	G 1 7	
E30	**Reinforcement for for in situ concrete**	G 5	Reinforcement
E30:1	Bar	G 5 1 - 4	
E30:1.1	Straight		)
E30:1 2	Bent		) No provision for
E30:1.3	Curved		) these classifications
E30:1.4	Links		)

SMM7		CESMM2	
Clause	Heading	Clause	Heading/Comment
E30:S1	Kind and quality of materials	G 5 1	Mild steel bars to BS4449
		G 5 2	High yield steel bars to BS4449 or BS4461
		G 5 3	Stainless steel bars of stated quality
		G 5 4	Reinforcing bars of other stated material
E30:2	Spacers and chairs	G 5 M8	
E30:3	Special joint	G 5 D7	
	Length classifications:-		
	Horizontal lengths 12.00 - 15.00m and thereafter in 3.00m stages	G 5 A7	Horizontal and vertical; lengths to next higher multiple of 3m where exceeding 12m before bending
	Vertical lengths 6.00 - 9.00m and thereafter in 3.00m stages		
E40	**Designed joints in in situ concrete**	G 6	Joints
E40:1	Plain	G 6 1	Open surface plain
E40:2	Formed	G 6 2	Open surface with filler
E40:3	Cut	G 6 3	Formed surface plain
	The above in m stating width or depth ≤ 150mm and thereafter in 150mm stages	G 6 4	Formed surface with filler
			The above in m2 - width or depth as follows:-
			not exceeding 0.5m 0.5 - 1m stated exceeding 1m

SMM7		CESMM2	
Clause	Heading	Clause	Heading/Comment
E40:4	Sealants in metres with dimensioned description	G 6 5	Plastics or rubber waterstops
		G 6 6	Metal waterstops
			The above in m - width as follows:-
			not exceeding 150mm
150 - 200mm
200 - 300mm
stated exceeding 300mm
E40:5	Angles in waterstops		deemed to be included - see Coverage Rule C4
E40:6	Intersections in waterstops		deemed to be included - see Coverage Rule C4
	Fillers, dowels included in descriptions - see Supplementary Information S2	)G 6 7
)
)
)	Sealed rebates or grooves
		)G 6 8	Dowels
			plain or greased
sleeved or capped
E41	**Worked finishes/
Cutting to
in situ concrete**	G 8	Concrete ancillaries
		G 8 1	Finishing of top surfaces
E41:1	Tamping by mechanical means		1 Wood float
2 Steel trowel
3 Other stated surface treatment
E41:2	Power floating		4 Granolithic finish
5 Other applied finish
E41:3	Trowelling		
		G 8 2	Finishing of formed surface

SMM7		CESMM2	
Clause	Heading	Clause	Heading/Comment
E41:4	Hacking		1 Aggregate exposure using retarder
E41:5	Grinding		2 Bush hammering
			3 Other stated surface treatment carried out after striking formwork
E41:6	Sandblasting		
E41:7	Finishings achieved by other means		
E41:8	Cutting chases	)	
E41:9	Cutting rebates	)	No rules provided
E41:10	Cutting mortices	)	
E41:11	Cutting holes	)	
E42	**Accessories cast into in situ concrete**		
E42:1	Type or name stated m2 m or nr	G 8 3	Inserts Linear inserts m Other inserts nr
		G 8 4	Grouting under plates
E50	**Precast concrete large units**	H	**PRECAST CONCRETE**
E50:1	Type or name stated nr or m stating nr floor units in m2 stating length	H 1	Beams)
		H 2	Pre-stressed) pre-tensioned) beams) nr
		H 3	Pre-stressed) post tensioned) beams)
		H 4	columns)
E50:2	Items extra over the units on which they occur Angles Fair ends Stoolings		Length and mass of the above to be stated in ranges
		H 5	Slabs Area and mass of slabs to be stated in ranges
		H 6	Segmental units

SMM7		CESMM2	
Clause	Heading	Clause	Heading/Comment
	Other details stated	H 7	Units for subways, culverts and ducts
		H 8	Copings, sills and weir blocks
E50:3	Joints		Joints deemed to be included - see Coverage Rule C1
E60	**Precast/Composite concrete decking**		
E60:1	Composite slabs		) No specific rules
E60:2	Formwork		) included
E60:3	Reinforcement		)

SMM7		CESMM2	
Clause	Heading	Clause	Heading/Comment
F	**MASONRY**	U	**BRICKWORK, BLOCKWORK AND MASONRY**
	The main sub divisions of this section are as follows:-		The corresponding sub divisions are as follows:-
F10	Brick/block walling	U 1 U 2 U 3 U 4 U 5	Common brickwork Facing brickwork Engineering brickwork Lightweight blockwork Dense concrete blockwork
F11	Glass block		
F20	Natural stone rubble walling	U 8	Rubble masonry
F21	Natural stone ashlar walling/ dressings	U 7	Ashlar masonry
F22	Cast stone walling/dressings	U 6	Artificial stone blockwork
	Walls classified as follows:-		Walls classified as follows:-
F10:1	Walls stating whether:-	U 1 1 1	Vertical straight walls
F10:1.1.1	Vertical	U 1 1 2	Vertical curved walls
F10:1.1.2	Battering	U 1 1 3	Battered straight walls
F10:1.1.3	Tapering one side	U 1 1 4	Battered curved walls
		U 1 1 5	Vertical facing to concrete
F10:1.1.4	Tapering both sides	U 1 1 6	Battered facing to concrete
		U 1 1 7	Casing to metal sections
F10:1.1.1.1	Built against other work		

SMM7		CESMM2	
Clause	Heading	Clause	Heading/Comment
F10:1.1 .1.4	Building overhand		
F10:1.1	Thickness stated		The above classified as to thickness as follows:- not exceeding 150mm -m2 150 - 200mm -m2 250 - 500mm -m2 500mm - 1m -m2 exceeding 1m -m3 the nominal thickness being stated - Additional Description Rule A5
	Walls include skins of hollow walls - Definition Rule D8		where the above are in cavity or composite construction this shall be stated Additional Description Rule A4
F10:1.2	Facework one side		Surface finish to be stated - Additional Description Rule A2
F10:1.3	Facework both sides	U 2 7 8	Surface features - fair facing
F10:2	Isolated piers	U 1 6	Columns and piers of stated cross-sectional dimensions - metres
F10:3	Isolated casings		
F20:3	Isolated columns		
F20:4	Attached columns		
F10:4	Chimney stacks		No specific provision made
F10:7	Isolated chimney shafts and the like		No specific provision made

SMM7		CESMM2	
Clause	Heading	Clause	Heading/Comment
F30:1	Forming cavities		No provision for this item but descriptions of walls to state if in cavity or composite construction - Additional Description Rules A4
F30:2	Damp proof courses - square metres	U 1 8 2	Damp proof course - metres
F30:3	Joint reinforcement	U 1 8 1	Joint reinforcement
F10:25	Bonding to existing in metres	U 1 8 4	Bonds to existing work in square metres

SMM7		CESMM2	
Clause	Heading	Clause	Heading/Comment
G	**STRUCTURAL/ CARCASSING/TIMBER**		
G10	**Structural steel framing**	M	**STRUCTURAL METALWORK**
G11	**Structural aluminium framing**		
G12	**Isolated structural metal members**	N	**MISCELLANEOUS METALWORK**
	Note: references below to G10 apply equally to G11 and G12		
	Fabrication classified as follows:-		Fabrication classified as - follows:-
G10:1.1	Columns	M 3 1	Columns
G10:1.2	Beams	M 3 2	Beams
G10:1.3	Bracings	M 3 6	Bracings, purlins and cladding rails
G10:1.4	Purlins and cladding rails		
G10.1.5	Grillages	M 3 7	Grillages
G10:1.7	Trestles, towers and built-up columns	M 3 4	Trestles, towers and built-up columns
G10:1.8	Trusses and built up girders	M 3 5	Trusses and built up girders
G10:1.9	Wires, cables, rods and bars	N 2 4	Tie rods
G10:1.10	Fittings		
			The above members are classsified as "members of frames" and "other members". Other divisions include:-

SMM7		CESMM2	
Clause	Heading	Clause	Heading/Comment
		M 3 3	Portal frames
G10:1:11	Holding down bolts or assemblies	M 3 8	Anchorages and holding down bolt assemblies
G10:2	Framing erection:-		
G10:2.1	Trial erection	M 5 1	Trial erection
G10:2.2	Permanent erection on site	M 5 2	Permanent erection
			The above two items are further classified as "members for bridges" "members for frames" and "other members"
G10:5	Isolated structural members	N 1 6	Miscellaneous framing
G10:7	Surface preparation	M 8	Off site surface treatment
G10:7.1	Blast cleaning	M 8 1	Blast cleaning
G10:7.2	Pickling	M 8 2	Pickling
G10:7.3	Wire brushing	M 8 4	Wire brushing
G10:7.4	Flame cleaning	M 8 3	Flame cleaning
G10:7.5	Others details stated		
G10:8	Surface treatment		
G10:8.1	Galvanizing	M 8 6	Galvanizing
G10:8.2	Sprayed metal coating	M 8 5	Metal spraying
G10:8.3	Protective painting	M 8 7	Painting
G10:8.4	Others details stated		

SMM7		CESMM2	
Clause	Heading	Clause	Heading/Comment
G10:9.1	Local protective coating - only applies to structural aluminum framing		
G20	**Carpentry/Timber framing/First fixing**	O	**TIMBER** Note: This class excludes Building carpentry and joinery
	Comparable classifications would be found with G20:1 to G20:13 inclusive	O 1	The divisions for this class are as follows:- Hardwood components
		O 2	Softwood components
			Classifications are given as to cross-sectional area and length
			The structural use and location of timber components is to be given for components longer than 3m
G20:20-28	Straps - connections etc.	O 5	Fittings and fastenings
G30	**Metal profiled sheet decking**	N	**MISCELLANEOUS METALWORK**
G30:1	Decking	N 1 1	Cladding
G30:2	Decking units		
G30:3	Holes notches etc on site or off site measured as extra over decking		No specific rules included
G30:4-.11	Bearings eaves etc. measured as linear items		

SMM7		CESMM2	
Clause	Heading	Clause	Heading/Comment
G31	<u>Prefabricated timber unit decking</u>	O	<u>TIMBER</u>
G31:1	Decking	O 3	Hardwood decking
G31:2	Decking units	O 4	Softwood decking
			Classifications are given for thickness
G31:3	Holes notches etc. on site or off site measured as extra over decking		No specific rules included

SMM7		CESMM2	
Clause	Heading	Clause	Heading/Comment
J	**WATERPROOFING**	W	**WATERPROOFING**
J20	Mastic asphalt tanking/damp proof membranes	W 1 1	Damp proofing asphalt
		W 2 1	Tanking asphalt
J21	Mastic asphalt roofing/ insulation/ finishes	W 3 1	Roofing asphalt
J30	Liquid applied tanking/damp proof membranes	W 1 4	Damp proofing waterproof coating
		W 2 4	Tanking waterproof coating
J31	Liquid applied waterproof roof coatings	W 3 4	Roofing waterproof coating
J32	Sprayed vapour	W 5	Sprayed or brushed waterproofing
J33	In situ glass reinforced plastics		
J40	Flexible sheet tanking/damp proof membranes	W 1 3	Damp proofing waterproof sheeting
		W 2 3	Tanking waterproof sheeting
J41	Built up felt roof coverings	W 3 3	Roofing waterproof sheeting
J42	Single layer plastics roof coverings	W 3 3	Roofing waterproof sheeting

SMM7		CESMM2	
Clause	Heading	Clause	Heading/Comment
M	**SURFACE FINISHES**		
M60	**Painting/clear finishing**	V	**PAINTING**
	The only classifications where comparable divisions are given are as follows:-		
M60:5	Structural metalwork	V 3 7	Metal sections
M60:9	Services	V 3 8	Pipework
	Supplementary information		
M60:S1	Kind and quality of materials	V 1	Lead, iron or zinc based primer paint
		V 2	Etch primer paint
		V 3	Oil paint
		V 4	Alkyd gloss paint
		V 5	Emulsion paint
		V 6	Cement paint
		V 7	Epoxy or polyurethane paint
		V 8	Bituminous or coal tar paint
M60:S2	Nature of base	V 1 1	Metal other than metal section and pipework
		V 1 2	Timber
		V 1 3	Smooth concrete
		V 1 4	Rough concrete
		V 1 5	Masonry
		V 1 6	Brickwork and blockwork
M60:S3	Preparatory work		Preparation deemed to be included - Coverage Rule C1
M60:S4-6	Number of coats		Number of coats or film thickness - Additional Description Rule A1

SMM7		CESMM2	
Clause	Heading	Clause	Heading/Comment
	Measurement classifications		Measurement classifications
M60:1	Girth > 300mm m2 Isolated surfaces girth ≤ 300mm m Isolated areas ≤ 0.5m2 irrespective of girth nr	V 1-8	Upper surfaces inclined at an angle not exceeding 30 degrees to the horizontal m2 Upper surfaces inclined at 30 degrees - 60 degrees to the horizontal m2 Surfaces inclined at an angle exceeding 60 degrees to the horizontal m2 Soffit surfaces and lower surfaces inclined at an angle not exceeding 60 degrees to the horizontal m2 Surfaces of width not exceeding 300mm m Surfaces of width 300mm - 1m m Metal sections and pipework m2

SMM7		CESMM2	
Clause	Heading	Clause	Heading/Comment
Q	PAVING/PLANTING FENCING/SITE FURNITURE	R	ROADS AND PAVINGS
Q10	Stone/Concrete/ Brick kerbs/ edgings/channels		
Q10: 2	Kerbs	R 6	Kerbs, channels and edgings
Q10:3	Edgings		
Q10:4	Channels		
Q10:5	Items extra over i.e. specials	R 6 1 3	Quadrants
		R 6 1 4	Drops
		R 6 1 5	Transitions
	Radius to be stated of curved items		Straight or curved radius exceeding 12m grouped together
			Curved to radius not exceeding 12m grouped together
Q21	In situ concrete roads/pavings bases		
		R 4	Concrete pavements
Q21:1	Concrete	R 4 3	Other in situ concrete slabs of state strength
Q21:2	Formwork		Deemed to be included see Coverage Rules C1
Q21:3	Reinforcement	R 4 4	Steel fabric reinforcement to BS4483
		R 4 5	Other fabric reinforcement
		R 4 6	Mild steel bar reinforcement to BS4449
		R 4 7	High yield steel bar reinforcement to BS4449 or BS4461

SMM7		CESMM2	
Clause	Heading	Clause	Heading/Comment
Q21:4	Joints	R 5	Joints in concrete pavements
Q21:5	Worked finishes		Finishes to concrete deemed to be included see Coverage Rules C1
Q21:6	Acessories cast in		
Q22	**Coated macadam/ Asphalt road/ pavings**		
Q22:1	Roads	R 2 2 R 2 3	Sub-bases, flexible roads bases and surfacing
Q22:2	Pavings		
Q22:3	Linings to channel	R 6 8	Asphalt channels
Q23	**Gravel/Hogging roads/pavings**	R	**Roads and Pavings**
Q23:1	Roads	R 1	Sub-bases, flexible road bases and surfacing
Q23:2	Pavings		
Q23:3	Edgings	R 6 4	In situ concrete kerbs and edgings
Q24	**Interlocking brick/block roads /pavings**	R 6 6	**Roads and Pavings**
Q24:1	Roads	R 4 2	Other carriageway slabs of stated strength
Q24:2	Pavings		
Q24:3	Treads		
Q24:4	Margins		
Q24:5	Risers		
Q24:6	Kerbs		
Q24:7	Edgings		

SMM7		CESMM2	
Clause	Heading	Clause	Heading/Comment
Q24:8	Linings to channels		
Q24:9	Items extra over the work in which they occur		
Q24:10	Accessories		
Q24:10.1	Separating membranes	R 4 8	Waterproof membranes below concrete pavements
Q24:10.2	Movement joints	R 5	Joints in concrete pavements
Q24:10.3	Tree grilles		
Q25	**Slab/Brick/Block /Sett/Cobble pavings**		
Q25:1	Roads	R 4 2	Other carriageway slabs of stated strength
Q25:2	Pavings	R 7 8	Precast concrete flags to stated specification
Q25:3	Treads		
Q25:4	Margins		
Q25:5	Risers		
Q25:6	Kerbs		
Q25:7	Edgings		
Q26:8	Linings to channels		
Q25:9	Items extra over the work in which they occur		
Q25:10	Accessories		

SMM7		CESMM2	
Clause	Heading	Clause	Heading/Comment
Q25:10.1	Separating membranes	R 4 8	Waterproof membranes below concrete pavements
Q25:10.2	Movement joints	R 5	Joints in concrete pavements
Q25:10.3	Tree grilles		
Q30	Seeding/Turfing	E	**EARTHWORKS**
Q30:1	Cultivating	E 8 C4	Trimming and preparation of surface deemed to be included
Q30:2	Surface applications	E 8 C4	Fertilizer is deemed to be included
Q30:3	Seeding	E 8 2	Hydraulic mulch grass seeding
		E 8 3	Other grass seeding
Q30:4	Turfing	E 8 1	Turfing
Q30:5	Turfing edges of seeded areas		
Q30:S5	Method of securing turves	E 8 A17	Pegging or wiring of turf to be stated
Q30:6	Protection - temporary fencing		
Q31	**Planting**		
Q31:3	Trees	E 8 6	Trees, stated species and size
Q31:4	Younger nursery stock trees	E 8 6	Trees, stated species and size
Q31:5	Shrubs	E 8 5	Shrubs, stated species and size
Q31:6	Hedge plants	E 8 7	Hedges, stated species, size and spacing

SMM7		CESMM2	
Clause	Heading	Clause	Heading/Comment
Q31:7	Herbaceous plants	E 8 4	Plants, stated species and size
Q31:8	Bulbs, corms and tubers		
Q31:9	Mulching after planting		
Q31:10	Protection		
Q40	**Fencing**	X	**MISCELLANEOUS WORK**
Q40:1	Fencing	X 1	Fences
Q40:2	Special supports extra over fencing in which they occur	X 1 C2	Fences shall be deemed to include end posts, straining posts and gate posts
Q40:3	Independent gate posts		
Q40:4	Items extra over fencing, special supports and independent gate posts irrespective of type		
Q40:5	Gates	X 2	Gates and stiles
Q40:6	Ironmongery		

SMM7		CESMM2	
Clause	Heading	Clause	Heading/Comment
R	DISPOSAL SYSTEMS	I J	PIPEWORK - PIPES PIPEWORK - FITTINGS VALVES
R10	Rainwater pipework/gutters		
R11	Foul drainage above ground		
	Note: References below to R10 apply equally to R11 where applicable		
R10:1	Pipes	I 1-8	Pipes are classified as "Not in trenches" or "In trenches"
		I 1 D1	Pipes not in trenches includes pipes suspended or supported above ground or other surface
R10:2	Items extra over the pipe in which they occur	J 1-8	Bends etc. not measured extra over the pipe
R10:3	Screwed sockets		
R10:4	Tappings		
R10:5	Bosses		
R10:6	Pipework ancillaries	J 8	Valves and penstocks
R10:7	Pipe supports which differ from those given with pipelines		
R10:8	Pipe sleeves through walls floors and ceilings		
R10:9	Wall, floor and ceiling plates		

SMM7		CESMM2	
Clause	Heading	Clause	Heading/Comment
R10:10	Gutters		No specific rules included
R10:11	Items extra over the gutter in which they occur		
R10:12	Marking position of holes, mortices and chases in the structure		
R10:13	Identification		
R10:14	Testing and commissioning		
R10:15	Temporary operation of installations to Employer's requirements		
R10:16	Preparing drawings		
R10:17	Operating and maintenance manuals		
R.12	<u>Drainage below ground</u>		
R12:1	Excavating trenches	I 1-8 C2	Items of "pipes in trenches" are deemed to include excavation, preparation of surfaces, disposal of excavated material, upholding sides of excavation, backfilling and removal of existing services except to the extent that such work is included in class J, K and L
R12:2	Items extra over excavating trenches, irrespective of depth		
R12:3	Disposal		
R12:4	Beds	L 3	Beds

SMM7		CESMM2	
Clause	Heading	Clause	Heading/Comment
R12:5	Beds and haunching	L 4	Haunches
R12:6	Beds and surrounds	L 5	Surrounds
R12:7	Vertical casings		
R12:8	Pipes	I 1-8	Pipes are classified as "Not in trenches" or "In trenches"
R12:9	Items extra over the pipes in which they occur	J 1 1-8	Bends etc. not measured extra over the pipe
R12:10	Pipe accessories	K 3	Gullies
R12:11	Manholes	K 1	Manholes
R12:12	Inspection chambers	K 2	Other stated chambers
R12:13	Soakaways	K 2	Other stated chambers
R12:14	Cesspits	K 2	Other stated chambers
R12:15	Septic tanks	K 2	Other stated chambers
R12:16	Connecting to Local Authority's sewer	K 8 6	Connections to existing pipes, ducts and culverts
R12:17	Testing and commissioning		
R12:18	Preparing drawings		
R12:19	Operating and maintenance manuals		
R13	Land drainage The classifications listed under R10 equally apply here where applicable	K 4	French drains, rubble drains, ditches and trenches

5 : POMI
– How it Compares with SMM7

The Royal Institution of Chartered Surveyors in 1979 published "Principles of Measurement (International)", a method of measurement prepared specifically for use internationally where circumstances, contracts, practices and techniques vary from country to country.

Principles of Measurement (International), commonly known as POMI, is a very concise method of measurement of only twenty-one pages inclusive of a blank page for writing amendments for use in a particular locality or for work not envisaged by the document.

Part of the brief of the SMM7 development unit was to simplify the method of measurement and thus reduce the number of items to be measured. To some extent this has been achieved. Many of the labours previously measurable in SMM6 are now deemed to be included. Even so, SMM7 is still some one hundred and ninety pages in length compared with the twenty-one of POMI.

The principles of measurement set out in POMI are open to more flexible interpretation than those of SMM7. It is therefore likely that Bills of Quantities measured in accordance with POMI could be disputed more easily than those measured in accordance with SMM7.

To assist those measuring on both methods of measurement we have prepared a comparison between the two documents. It was not our intention to compile this comparison in the same detail as that prepared to compare SMM7 with SMM6 but we have endeavoured to highlight the major differences.

The comparison which follows has been set out in SMM7 Work Section Order. To assist in finding particular POMI clause references a list of POMI items in alpha/numeric order with page numbers is included as Appendix C at the end of this book.

SMM7		POMI	
Clause	Heading	Clause	Heading/Comment
	General Rules	GP	**General Principles**
1	Uniform basis for measuring building works	GP1	Very similar wording
2	Tabulated format		POMI is set out in prose as previous methods of measurement
		GP2	Bills of Quantities. An explanation of what they should contain and of what should be provided with them
3	Quantities. Measured net as fixed. To nearest 10mm. Tonnes to two decimal places. Explanation of minimum deducts for voids	GP3 GP3.5	Measurement. Very similar - no deduction for voids less than 1.0sq.m unless otherwise stated. A separate minor building or structure can be enumerated
4	Descriptions	GP5	Description of items
5	Drawn information. Sets out the type of drawn information which should be available and accompany the Bills of Quantities	GP2.3	Drawings shall be provided with the Bill of Quantities. No other specific requirements are stated
6	Catalogued or standard components. A precise reference from a catalogue is sufficient information	GP5.5 and 5.6	Similar rules

SMM7		POMI	
Clause	Heading	Clause	Heading/Comment
7	Work of special types. (Existing buildings, temporary works, work outside curtilage, work in or under water)	–	Not stated but such headings as are suggested in SMM7 would probably be automatically used in POMI to emphasis the special nature of the work
8	Backgrounds. Definitions of special backgrounds. Fixing to these has to be identified	–	Not mentioned
9	Composite items	F7	Composite items. Very similar definition
10	Procedure where drawn and specification information not available. Explanation of when to use approximate quantities and/or provisional sums	A2.1	Not really applicable to POMI as it does not rely on the provision of detailed drawn information. Where the specification contains clauses related to any of the General Requirements (A3-A9), the Bills of Quantities shall make reference to the apppropriate clauses
11	Work not covered. Rules to be construed	GP1.2	The rules used are to be stated. A page is provided at the end of the document for this purpose
12	Symbols and abbreviations - a full list of all those used in the document with their meanings	–	Symbols are not used
13	Work to Existing Buildings	–	Not mentioned but would be stated as suggested in rule GP1.1

	SMM7		POMI
Clause	Heading	Clause	Heading/Comment
14	General definitions. Definition of "curved, radii stated"	–	Not given

	SMM7		POMI	
Clause	Heading	Clause	Heading/Comment	
A	**PRELIMINARIES/ GENERAL CONDITIONS**	A	**GENERAL REQUIREMENTS**	
	The requirements, whilst being in a different order, are very similar in both documents. The main difference is the requirement in SMM7 for various items to be split into fixed and time related charges. SMM7 gives more examples of items which should be included but an experienced surveyor would use very similar clauses if he were working with POMI			
A51,	Nominated Sub-Contracts	GP6 to 9	In POMI these are dealt with in General Principles	
A52	Nominated Suppliers			
A53	Work by Statutory Authorities			
A55	Dayworks			

	SMM7		POMI
Clause	Heading	Clause	Heading/Comment
B	**COMPLETE BUILDINGS**		
	No specific rules	–	As SMM7

SMM7		POMI	
Clause	Heading	Clause	Heading/Comment
C	DEMOLITION/ ALTERATION/ RENOVATION	B5 B6	DEMOLITIONS AND ALTERATIONS SHORING
C10	Demolishing structures		
C10:1	Demolishing all structures	)))	Demolishing all or individual structures to be given as items.
C10:2	Demolishing individual structures	)B5.3)))	More detail requested in SMM7 but very similar detail would need to be given using
C10:3	Demolishing parts of structures	))	POMI even though not specifically requested
C30	Shoring		
C30:C1b		B6.1	Shoring incidental to demolitions
C30:4-5	Support of structures not to be demolished and roads	B6.2 to .4	Comment as above for B5.3
C10:6-7	Temporary roofs Temporary screens	B5.5	Temporary screens and roofs
C20	Alterations - spot items	B5.2 and B5.4	Removing individual fittings Cutting openings and
C20:1-8. *.1	Making good to be stated		alterations to existing structures - making good is understood to be included
C20:9-10	Temporary roofs Temporary screens	B5.5	Temporary screens and roofs

SMM7		POMI	
Clause	Heading	Clause	Heading/Comment
C40	<u>Repairing/ renovating concrete/ brick/block/ stone</u>	))))))	
C41	<u>Chemical dpc's to existing walls</u>	)))	- Not specifically mentioned in POMI
C50	<u>Repairing/ Renovating metal</u>	)))	
C51	<u>Repairing/ Renovating timber</u>	)))	
C52	<u>Fungus/Beetle eradication</u>	))	

SMM7		POMI	
Clause	Heading	Clause	Heading/Comment
D	**GROUNDWORK**	B	**SITEWORK**
D20	**Excavating and filling**		
P1	Information required to be given:-		
P1(a) to (c)	Ground water level	B1 & B8.1	Details shall be provided of all known
P1(d)	Trial pit details	B2 & B3	features, ground
P1(e)	Retained features		conditions, strata
P1(f)	Live over or underground services		etc.
P1(g)	Pile sizes and layout (D30-32) where applicable	GP5.6	Descriptions of items in Bills of Quantities may refer to other documents or drawings
D20:1	Site preparation	B4	No requirement to state girth of trees. Removal of hedges is covered in SMM7 as part of "Clearing site vegetation"
D20:2	Excavating		
D20:2.1	Topsoil for preservation - superficial measure	B9.1.1	Removal of topsoil to be measured by volume stating average depth
D20:2.2 to 8	Reduced level, basement, pit, trench excavation	B9.1.2-6	All as SMM7 except B9.1.3 "Excavation in cuttings" which is not mentioned in SMM7
D20:3	Excavating .1 Below ground water level .2 Next existing services .3 Around existing services	-	Not specifically mentioned but full details of all known problem areas have to be given

SMM7		POMI	
Clause	Heading	Clause	Heading/Comment
D20:4-5	Breaking out various materials given as "Extra over" excavation	B8.5	Excavation in rock can be so described or measured "Extra over" excavation.
		B8.6	The definition of rock is very similar to that in SMM7 except it is the "employer's representative" who decides whether the material is to be classed as rock. No mention is made of breaking up or breaking out other materials but allowance would need to be made
D20:6	Working space. Clearly defined instances when this is measurable	B8.2	No allowance to be made for any working space. Not a measurable item
D20:7	Earthwork support - to be measured in stages - maxmimum depth not exceeding 1.00m - - maximum depth not exceeding 2.00m - thereafter in 2.00m stages in square metres with various measurement and definition rules	B8.4	Earthwork support to be given as an item. The Contractor will therefore have to assess the amount required from the various volumes of excavation
D20:8.1 & .2	Disposal of water	A8.1.5 & 6	Pumping and de-watering are dealt with as temporary works in "General Requirements"
D20:8.3	Disposal of excavated materials	B11	Disposal - no difference

SMM7		POMI	
Clause	Heading	Clause	Heading/Comment
D20/Q20 :9&:10	Filling to excavations Filing to make up levels	) B12)))	Filling - no difference
D20/Q20 :11	Filling to external planters	-	Not mentioned
D20:12 & :13	Surface packing Surface treatments	) -))	Not mentioned
D30	<u>Cast in place piling</u>	B13-18 B15	Piling Bored piling
D31	<u>Preformed concrete piling</u>	)B14))	Driven piling
D32	<u>Steel piling</u>	)B16	Sheet piling
D40	<u>Diaphragm walling</u> A separate section covering all aspects of diaphragm walling	B9.7 C2.1.10	Diaphragm walling is to be measured under the various section heads
D50	<u>Underpinning</u>. A separate section covering all aspects of underpinning	B7	Underpinning - excavation, temporary supports and cutting away projecting foundations. Other work to be measured in accordance with the appropriate sections

SMM7		POMI	
Clause	Heading	Clause	Heading/Comment
E	**IN-SITU CONCRETE LARGE PRECAST CONCRETE**	C	**CONCRETE WORK**
E10	**In-situ concrete**		
E10:1-3	Foundations and isolated foundations include attached and isolated pile caps and column bases respectively Ground beams are measured separately	C2.1.1	Foundations include column bases but pile caps, which include ground beams, are measured separately
E10:4	Beds are measured in thickness stages and include blinding	C2.1.4	Beds should state actual thickness. Blinding is measured separately
E10:5	Slabs are measured in thickness stages and include attached beams	C2.1.5 C2.1.8	Suspended slabs include landings but attached beams are measured separately
E10:13	Landings are included with staircases		
E10:6	Coffered and troughed slabs	C2.2	Suspended slabs of special construction
E10:7	Walls - measured cubic - thickness in stages	C2.1.6	Walls - include attached columns - thickness to be stated - measured cubic
E10:8	Filling to hollow walls	D5.1	Concrete filling to cavities - measure by area

SMM7		POMI	
Clause	Heading	Clause	Heading/Comment
E10:9 :10	Beams Beam casings Attached beams where depth is not exceeding 3 times their width are measured with slabs	C2.1.8	Beams. All beams are measured separately
E10:11 & :12	Columns Column casings	C2.1.7	Columns - no difference
E10:13	Staircases - includes landings and strings	C2.1.9	Staircases - includes strings but landings measured with suspended slabs (C2.1.5)
E10:14	Upstands	-	Measure with Beds or Suspended slabs as appropriate
E10:15	Extra over item .1 Working around heating pipes .2 Monolithic finishes		Not mentioned. Use discretion. Refer to in description or measured extra over if considered appropriate
E10:16	Grouting - enumerated	E2.4	Wedging up and grouting bases to be enumerated
E10:17	Filling - mortices holes etc.	C7.7	Mortices and holes are understood to be included
E11	**Gun applied concrete**	C1.2	Poured concrete required by the specification to be placed, compacted, cured or otherwise treated in a particular manner shall be so described

SMM7		POMI	
Clause	Heading	Clause	Heading/Comment
E20	**Formwork**	C4	**Shuttering**
			Rules for the various types of shuttering are all very similar to those for formwork in SMM7 except:
E20:1	Sides of foundations	)))	
E20:2	Sides of ground beams	)C4.1.4)	Sides of foundations include sides of ground beams
E20:9	Soffits of landings	C4.1.1	Soffits include "soffits of landings"
E20:15 & :16	Columns Column casings	C4.1.5	Shuttering to attached columns is included with walls
E20:25	Stair flights include formwork to soffits	C4.1.2	Sloping soffits include soffits of staircases
E30	**Reinforcement**	C3	**Reinforcement**
			No significant differences but there is no requirement to state the length of long bars
E31	**Post tensioned reinforcement for in-situ concrete**	–	Not specifically mentioned but details would need to be given where applicable
E40	**Designed joints in in-situ concrete**	C7.4	Designed joints. SMM7 asks for more detailed information but the same information should be given when measuring under POMI
E41	**Worked finishes/ cutting to in-situ concrete**	C7	Sundries

SMM7		POMI	
Clause	Heading	Clause	Heading/Comment
E42	<u>Accessories cast into in-situ concrete</u>	C7.6	Fixings, ties, inserts or the like are to be enumerated or measured by area
E50	<u>Precast concrete large unit</u>	)C5)C6))	<u>Precast concrete</u> <u>Prestressed concrete</u>
E60	<u>Precast/Composite concrete decking</u>	))	
			Reinforcement to be measured separately under a heading in accordance with C3

SMM7		POMI	
Clause	Heading	Clause	Heading/Comment
F	**MASONRY**	D	**MASONRY**
F10	**Brick/block walling**		
F11	**Glass block walling**		
F10:1	Walls	D2.1.1	Walls. Piers within walls are measured as walls of combined pier and wall thickness
	Include skins of hollow walls (D4)	D2.1.3	Cavity walls measured as composite item of skins and cavity
F10:2	Isolated piers	)D2.1.4	Isolated piers
F10:3	Isolated casings	)	
F10:4	Chimney stacks	)	
F10:5	Projections	D2.1.1	Measure as walls of combined wall and pier thickness
F10:6	Arches	D3.2	Arches
F10:7	Isolated chimney shafts	(D2.1.4)	Not mentioned. Measure as isolated pier (D2.1.4)
F10:8	Boiler seatings	Q9	Work incidental to mechanical engineering installations
F10:9	Flue linings		
F10:10	Boiler seating kerbs		
F10:11	Items extra over	–	Understood to be included with item of walls
	.1.1 Reveals		
	.1.2 Intersections		
	.1.3 Angles		
F10:12	Closing cavities	D2.1.3	Cavity walls. Closing cavities understood to be included
F10:13	Facework, ornamental bands and the like	D3.1	Oversailing and receding courses measured by length

SMM7		POMI	
Clause	Heading	Clause	Heading/Comment
F10:14	Facework quoins	)D3.1	Sills, copings and the
F10:15	Facework sills	)	measured by length
F10:16	Facework thresholds	))	
F10.17	Facework copings	)	
F10:18	Facework steps	)	
F10:19	Facework tumblings to buttresses	))	
F10:20	Facework key blocks	))	
F10:21	Facework corbels	)	
F10:22	Facework bases to pilasters	))	No specific rules
F10:23	Facework cappings to pilasters	))	
F10:24	Facework cappings to isolated piers	))	
F10:25	Bonding to existing	-	Not mentioned
F10:26	Surface treatments	))	No specific rules
F20	<u>Natural stone rubble walling</u>	))	D	No specific rules but the rules for Masonry
F21	<u>Natural stone/ ashlar walling/ dressings</u>	)))	will apply. The rules for masonry are limited when they are
F22	<u>Cast stone walling/dressings</u>	))	to be used to measure natural or cast stone. Additional rules will need to be considered
F30	<u>Accessories/ Sundry items for brick/block/stone walling</u>		
F30:1	Forming cavities	D2.1.3	Cavity walls. Forming cavities in hollow walls can be included in the composite item of the cavity walls or the skins and cavity can each be measured separately. Forming other cavities would be measured separately

| \multicolumn{2}{c|}{SMM7} | \multicolumn{2}{c}{POMI} |

Clause	Heading	Clause	Heading/Comment
F30:2	Damp proof courses	G3	Damp proof courses
F30:3	Joint reinforcement	D4	Reinforcement to be measured in accordance with C3. Alternatively fabric reinforcement may be measured by length
F30:4	Weather fillets	)	
F30:5	Angle fillets	)	
F30:6	Pointing in flashings	) -	No specific rules
F30:7	Wedging and pinning	)	
F30:8	Joints	D5.2	Expansion joints or the like
F30:9	Slates and tiles for creasing	)D3	Sills - measure by length and describe
F30:10	Slate and tile cills	)	
F30:11	Flue linings	Q9	Work incidental to mechanical engineering installations
F30:12	Air bricks	)D5.3	Air bricks or the like to be enumerated. Include opening etc. in the description
F30:13	Ventilating gratings	)	
F30:14	Soot doors	)	
F30:15	Gas flue block	Q9	Work incidental to mechanical engineering installations
F30:16	Proprietary items	GP5.5	Notwithstanding the principles of measurement, proprietary items may be measured in a manner appropriate to the manufacturer's tariff or practice
F31	<u>Precast concrete sills/lintels/ copings/features</u>	C5 C6	<u>Precast concrete</u> <u>Prestressed concrete</u>

SMM7		POMI	
Clause	Heading	Clause	Heading/Comment
G	**STRUCTURAL/ CARCASSING METAL/ TIMBER**	E	**METALWORK**
G10	<u>Structural steel framing</u>	E2	Structural metalwork
G11	<u>Structural aluminium framing</u>		Less detail given than in SMM7 but general rules for measurement are very similar
G12	<u>Isolated Structural members</u>		
G10:M2	Fittings only measured separately where of a different type and grade of material	E2.2	Fittings to be given as an item
G10:C1	Shop and site black bolts are deemed to be included	E2.3	Fixings to be given as an item
G10:8 & G10:9	Surface treatment Localised protective coating - measured in square metres	E2.6	Protective treatments to be given as an item
G20	<u>Carpentry/Timber Framing/First Fixing</u>	F F2 F5	**WOODWORK** Structural members Framework
G20:1	Trusses	)F7	Composite items
G20:2	Trussed rafters	)	
G20:3	Trussed beams	)	
G20:4	Wall or partition panels	)	
G20:5	Portal frames	)	
G20:6	Floor members	)	
G20:7	Wall or partition members	)F2.1)	Structural timber measured under similar headings as SMM7
G20:8	Plates	)	
G20:9	Roof members	)	
G20:10	Joist strutting	F2.2	No difference

SMM7		POMI	
Clause	Heading	Clause	Heading/Comment
G20:11	Butt-jointed supports	)F4)	Grounds and battens. Firrings and bearers understood to be included with boarding to roof (F3.1.4)
G20:12	Framed support	)	
G20:13	Individual support	))	
G20:14	Gutter board	)	
G20:15	Fascia board	)	
G20:16	Eaves or verge soffit board	)F3)	Boarding and flooring
G20:17	Cleats	)	
G20:18	Ornamental ends	)	
G20:19	Wrot surfaces	F8.1	Finished surfaces on sawn items
G20:20 to :28	Metalwork associated with timber	F9.1	Metalwork
G30	**Metal profiled sheet decking**	G2	Coverings and linings. Very similar, though less detailed, rules
G31	**Prefabricated timber unit decking**	F7 G2	Composite items or Coverings and linings
G32	**Edge supported/ reinforced woodwool slab decking**		

SMM7		POMI	
Clause	Heading	Clause	Heading/Comment
H	CLADDING/COVERING		
H10	Patent glazing	H6	Patent glazing
H12	Plastics glazed vaulting/walling	))	Not specifically mentioned but measure
H13	Structural glass assemblies	))	in accordance with H6
H11	Curtain walling	H3.1	Screens, borrowed lights, curtain walling to be measured by area or alternatively enumerated
H14	Concrete rooflights/ pavement lights	C5 F7	Not specifically mentioned but could be measured under precast concrete or Composite items
H20	Rigid sheet cladding	G2 G4	Coverings and Linings Insulation - can be included with the description of the covering. In SMM7 insulation if not part of a roof decking is measured in detail in accordance with section P10 which covers sheet, quilt, board and loose fill insulations. Note:- G2 - Coverings and linings covers many different materials and types of construction and the rules for measurement will have to be adapted as appropriate

SMM7		POMI	
Clause	Heading	Clause	Heading/Comment
H21	Timber weatherboarding	)F3)OR	Boarding and flooring
K11	Rigid sheet flooring/sheating /linings/casings	)F6.7))	Sheet linings to walls and ceilings
K12	Under purlins/ Inside rail panel linings	)))	
K13	Rigid sheet fine linings/panelling	))	
K20	Timber board flooring/sheating/ linings/casing	)))	
K21	Timber narrow strip flooring/ linings	)))	
H30	Fibre cement profiled sheet cladding/covering/ siding	)G2)G4))	Refer to comments against SMM7 Clause H20 above
H31	Metal profiled/ flat sheet cladding/covering/ siding	))))	
H32	Plastics profiled sheet cladding/ covering/siding	)))	
H33	Bitumen and fibre profiled sheet cladding/covering	)))	
H41	Glass reinforced plastics cladding/ features	)))	
H40	Glass reinforced cement cladding/ features	C5 C6	Precast concrete Prestressed concrete
H50	Precast concrete slab cladding/ features	C5 C6	Precast concrete Prestressed concrete
H51	Natural stone slab cladding/ features	)J))	Finishes - measure in square metres giving full details
H52	Cast stone slab cladding/features	))	of joints, fixings etc.

	SMM7		POMI	
Clause	Heading	Clause	Heading/Comment	
H60	Clay/concrete roof tiling	)G2)G4	Refer to comments against SMM7 clause H20 above	
H61	Fibre cement slating	))		
H62	Natural slating	)		
H63	Reconstructed stone slating/ tiling	)))		
H64	Timber shingling	)		
H70	Malleable metal sheet prebonded coverings/ cladding	))))		
H71	Lead sheet coverings/ flashings	)))		
H72	Aluminium sheet coverings/ flashings	)))		
H73	Copper sheet coverings/ flashings	)))		
H74	Zinc sheet coverings/ flashings	)))		
H75	Stainless steel sheet coverings/ flashings	)))		
H76	Fibre bitumen thermoplastic sheet coverings/ flashings	))))		

SMM7		POMI	
Clause	Heading	Clause	Heading/Comment
J	**WATERPROOFING**	G	**THERMAL AND MOISTURE PROTECTION**
J10	Specialist waterproof rendering	J2.2	All walls, floors and ceilings measured in square metres irrespective of width
J20	Mastic asphalt tanking/damp proof membranes	)G2)G4)	Refer to comments against SMM7 Clause H20 above
J21	Mastic asphalt roofing/insulation/finishes	))))	
	Proprietary roof decking with asphalt finish	)))	
J30	Liquid applied tanking/damp proof membranes	)))	
J31	Liquid applied waterproof roof coatings	)))	
M11	Mastic asphalt flooring	)))	
J40	Flexible sheet tanking/damp proof membranes	)G2)G4)	
J41	Built up felt roof coverings	))	
J42	Single layer plastic roof coverings	)))	
J43	Proprietary roof decking with felt finish	)))	

SMM7		POMI	
Clause	Heading	Clause	Heading/Comment
K	**LININGS/SHEATHING DRY PARTITIONING**	J K	**FINISHES** **ACCESSORIES**
K10	**Plasterboard dry lining**		
K31	**Plasterboard fixed partitions/inner wall/linings**		
	Note: References below to K10 apply equally to K31		
K10:1	Proprietary paritions	K2	Partitions
K10:2	Linings. Reveals etc. measured in stages - not exceeding 300 - 300 - 600mm	J3	Finishings - reveals, returns, recesses, attached and unattached columns all to be included with walls
K10:3-8	Labours in forming angles, abutments etc.	-	Included with measurement of lining
K10:9	Beads - measured in linear metres with full descriptions	J4	Sundries. Beads etc. - measure as SMM7 but no requirement to state horizontal or vertical
K11,12, 13,20 & 21	See under H21 above		
K30	**Demountable partitions**	K	**Accessories** - partitions are measured under K2.1 by length over all doors and glazed units

SMM7		POMI	
Clause	Heading	Clause	Heading/Comment
K30:3	Doors, windows, etc. within demountable partitions are measured extra over the partitions and are deemed to include ironmongery, glass and linings but exclude trim	K2.2	Door and glazed units are enumerated stating the partition in which they occur (ie. extra over the partition as in SMM7)
K32	<u>Framed panel cubicle partitions</u>	K2.3	Cubicles
K33	<u>Concrete/Terrazzo partitions</u>	C5 C6	<u>Precast concrete</u> <u>Prestressed concrete</u>
K40	<u>Suspended ceilings</u>	J5	Suspended ceilings. Very much more detailed information required to be given in SMM7. Edge trims, angle trims and the like are all measured separately. POMI requires far less information to be given but much of the information required by SMM7 would be given by the competent surveyor as many of these items will have a substantial effect on price
K41	<u>Raised access floors</u>	J3.1.1	Not specifically mentioned but measured under Floors

SMM7		POMI	
Clause	Heading	Clause	Heading/Comment
L	**WINDOWS/DOORS/ STAIRS**	H F	**DOORS AND WINDOWS WOODWORK**
L10	**Timber windows/ rooflights/ screens/louvres**	H2 H3	Windows Screens
L11	**Metal windows/ rooflights/ screens/louvres**	H4	Ironmongery
L12	**Plastic windows/ rooflights/ screens/louvres**		
	All items are enumerated	H2.1 H3.1	Windows and skylights are enumerated Screens, borrowed lights and curtain walling are measured by area or enumerated
L10:8 - :10	Bedding and pointing frames	–	Bedding and pointing not mentioned but could be referred to in the description of item
L20	**Timber doors/ shutters/hatches**	)H1)H4	Doors Ironmongery
L21	**Metal doors/ shutters/ hatches**	)))	
L22	**Plastics/Rubber doors/Shutters/ hatches**	)))	
L20:8 - :10	Bedding and pointing frames	–	Bedding and pointing not mentioned but could be referred to in the description of item
L30	**Timber stairs/ walkways/ balustrades**	F7	Woodwork - Composite items
L31	**Metal stairs walkways/ balustrades**	E3.3	Metalwork - Non structural

SMM7		POMI	
Clause	Heading	Clause	Heading/Comment
L40	<u>General glazing</u>	H5	Glass is measured by area in all cases except sealed factory - glazed units, louvres and panes of special shape or with decorative treatment which are enumerated. Louvres may be measured by length, stating the number
L41	<u>Lead light glazing</u>		
L42	<u>Infill panels/ sheets</u>		

SMM7		POMI	
Clause	Heading	Clause	Heading/Comment
M	SURFACE FINISHES	J	FINISHES
M10	Sand cement/ Concrete/ Granolithic screeds/flooring	J2.2	All walls, floors and ceilings measured in square metres irrespective of width
M12	Trowelled bitumen/ resin/rubber-latex flooring		
M20	Plastered/ Rendered/Roughcast coatings		
M23	Resin bound mineral coatings		
J10	See Section J		
	N.B. References to M10 apply equally to M12, M20 & M23		
M10:3 & :4	Beams Columns	J3.1	Beams and columns are measured with ceilings and walls respectively not separately
M10:7 M10:8 M10:9 M10:10	Treads Risers Strings Aprons	)J3.1.4)))	Staircases include work to treads, risers and edges of landings
M11	Mastic asphalt floor See under J31 above		
M21 M22	Insulation with rendered finish Sprayed mineral fibre coatings	)J3)))	Finishings - All work to walls, floors, ceilings and staircases measured in square metres irrespective of width. Beams and columns are measured with ceilings and walls respectively and not separately

SMM7		POMI	
Clause	Heading	Clause	Heading/Comment
M30	<u>Metal mesh reinforcement/ Anchored reinforcement for plastered coatings</u>	J2.3	Pre-formed backgrounds. All work to walls, floors and ceilings, measured in square metres irrespective of width
M31	<u>Fibrous plaster</u>	J3	Finishings
M40	<u>Stone/Concrete/ Quarry/Ceramic tiling/Mosaic</u>	)J3)J2.3)	Finishings Pre-formed backgrounds All work to walls,
M42	<u>Wood block/ Composition block/ Parquet flooring</u>	)))	floors and ceilings measured in square metres irrespective
M41	<u>Terrazzo tiling/ insitu terrazzo</u>	))J3.1	of width Beams and columns
M50	<u>Rubber/Plastics/ Cork/Lino/Carpet tiling/sheeting</u>	)))	are measured with ceilings and walls respectively and
M51	<u>Edge fixed carpeting</u>	))	not separately
M52	<u>Decorative papers fabrics</u>	J6	Decorations. Papers/ fabrics not specifically mentioned but to be measured in square metres
M60	<u>Painting/Clear finishing</u>	J6	Decorations to general surfaces, frames, windows, floors, walls, ceilings, radiators, gutters, large pipes and structural metalwork are all measured by area. No distinction is made for narrow widths. Decoration on small pipes (not exceeding 60mm internal diameter) are measured in linear metres Decoration on gratings, rainwater heads or the like are enumerated

	SMM7		POMI
Clause	Heading	Clause	Heading/Comment
M60:7	Decorations to fences, gates, etc.	B21.4	Decoration to be measured in accordace with Section J

SMM7		POMI	
Clause	Heading	Clause	Heading/Comment
N	**FURNITURE/ EQUIPMENT**	L M	**EQUIUPMENT FURNISHINGS**
N11	**Domestic kitchen fittings**		By the very nature of the components
N12	**Catering equipment**		being measured both
N13	**Sanitary appliances/ fittings**		methods of measurement require them to be
N15	**Signs/Notices**		enumerated giving
N20-23	**Special purpose fixtures/ furnishings/ equipment**		full description
Q50	**Site/Street furniture/ equipment**		
N13:1	Sanitary fittings	Q4.1	Equipment in connection with mechanical engineering installations
N15:2	Signwriting	J7	Signwriting

SMM7		POMI	
Clause	Heading	Clause	Heading/Comment
P	**BUILDING FABRIC SUNDRIES**		
P10	<u>Sundry insulation/ proofing work/ fire stops</u>	)G4)))	Insulation
P11	<u>Foamed/Fibre/Bead cavity wall insulation</u>	)))	
P20	<u>Unframed isolated trims/skirtings/ sundry items</u> All components are to be itemised separately	F	**Woodwork**
P20:1	Skirtings, picture rails, architraves and the like	F.6.1	Finishings and fittings - cover fillets, architraves, skirtings, beads stops, edgings, window boards, nosings or the like. Architraves and skirtings can be grouped together as cover fillets (F6.1.1). Stops are grouped with beads (F6.1.2). Nosings are measured as edgings (F6.1.3)
P20:2	Cover fillets, stops, trims, beads, nosings and the like		
P20:3	Isolated shelves and worktops	F6.2 F6.3	Worktops, seats or the like
P20:4	Window board	F6.1.3	Edgings which shall include window boards
P20:7	Handrails	F6.2.2	Handrails or balustrades
P20:8	Hardwood members over 0.003 m2 sectional area are to have ends, angles, mitres and intersections measured	-	No similar requirement in POMI

SMM7		POMI	
Clause	Heading	Clause	Heading/Comment
P20:9	Backboards, plinth blocks	F6.4	Backboards
P21	<u>Ironmongery</u> (not ironmongery supplied with components - L10.Cl(c))	F10	Ironmongery
P22	<u>Sealant joints</u>	-	Not specifically mentioned but could be measured under J4 Sundries on Finishings
P30	<u>Trenches/ Pipeways/Pits for buried engineering services</u>	)Q9)& R10)))	Work incidental to services installation - to be measured in accordance with
P31	<u>Holes/Chases/ Covers/Supports for services</u>	)))	P3 and the relevant sections of the document and grouped under a heading

SMM7		POMI	
Clause	Heading	Clause	Heading/Comment
Q	**PAVING/PLANTING FENCING/SITE FURNITURE**	B	**SITE WORKS**
Q10	**Stone/Concrete/ Brick/kerbs/ edgings/channels** - measured by lengths - full details to be stated	B20.3	Channels, curbs, edgings - no difference
Q21	**In-situ concrete roads/pavings/ bases** - items of concrete formwork, reinforcement etc. are all measured separately in accordance with the relevant section	B20.1	Paving and surfacing to be measured by area - specification etc. can be given in description
Q22	**Coated macadam/ Asphalt roads/ pavings**	)B20.1)	Paving and surfacing
Q23	**Gravel/hoggin roads/pavings**	))	
Q24	**Interlocking brick/block roads/ pavings**	)))	
Q25	**Slab/Brick/Sett/ Cobble pavings**	))	
Q26	**Special surfacings pavings for sport**	))	
Q30	**Seeding/Turfing**	B22.1 & B22.2	Landscpaing - Cultivating and fertilising. Sorting, seeding and turfing
Q31	**Planting**	B22.3 & B22.4	Hedges Trees and shrubs

SMM7		POMI	
Clause	Heading	Clause	Heading/Comment
Q40	**Fencing**		
Q40:1	Fencing	B21.1	Fencing
Q40:2	Special supports	B21.2	Special posts
Q40:3	Independant gate posts		
Q40:5	Gates	B21.3	Gates, barriers or the like
Q40:6	Ironmongery - measured in accordance with P21	-	Not mentioned but could be included with description of gate or measured in accordance with F10

\	SMM7	\	POMI
Clause	Heading	Clause	Heading/Comment
R	DISPOSAL SYSTEMS	Q	MECHANICAL ENGINEERING INSTALLATIONS
R10	Rainwater pipework/gutters Foul drainage above ground	Q1.2	Installation may be measured in detail or enumerated on a locational basis
R10:1	Pipes	Q2.1	Pipes and gutters
R10:2	Fittings to be measured extra over with details of each type of fitting	Q2.2	Fittings to small pipes (not exceeding 60mm) are understood to be included. Fittings to large pipes are enumerated for each size of pipe but grouped together as "fittings"
R10:6	Ancillaries all enumerated	Q2.3	Valves, traps enumerated
R10:7	Pipe supports which differ from those given with pipelines	–	Not specifically mentioned
R10:8	Pipe sleeves – enumerated	Q2.3	Sleeves understood to included
R10:9	Wall, floor and ceiling plates	Q2.3	Cover plates understood to be included
R10:10	Gutters	Q2.1	Gutters
R10:11	Special joints and fittings to gutters enumerated extra over the gutters. Full details given	Q2.2	Fittings etc. enumerated but grouped together as "fittings"
R10:12	Marking positions to be given as an item	Q9	Work incidental to mechanical installation - referred to P3

SMM7		POMI	
Clause	Heading	Clause	Heading/Comment
R10:13	Identification	)Q8	Sundries - referred to P2
R10:14	Testing and commissioning	)	
R10:15	Temporary operation of installation to Employer's requirements	)	
R10:16	Preparing drawings	)	
R10:17	Operating and maintenance manuals	)	
R12	**Drainage below ground**	)B19	Underground drainage
R13	**Land drainage**	)	
R12:1	Excavating trenches	B9.2	Excavation of trenches - disposal and filling understood to be included
R12:3	Disposal of water - to be given as items for "surface" and "ground" water	A8.1.6	De-watering. To be given in "General Requirements" as an item of "temporary works"
R12:4	Beds	)B19.4	Beds and coverings - formwork to be stated if required. Surrounds to vertical-pipes formwork understood to be included
R12:5	Beds and haunchings	)	
R12:6	Beds and surrounds	)	
R12:7	Vertical casings	)	
R12:8	Pipes - iron pipes not exceeding 3m long to be measured in metres stating the number	B19.1	Drain pipes - no requirement to differentiate lengths not exceeding 3m
R12:9	Pipe fittings - to be enumerated with detailed description	B19.2	Drain fittings - to be enumerated, grouped together for each size of pipe and given as "fittings"

SMM7		POMI	
Clause	Heading	Clause	Heading/Comment
R12:10	Pipe accessories	B19.3	Drain accessories - no difference
R12:11	Manholes	)B19.5	Inspection chambers to be enumerated or measured in accordance with the relevant sections
R12:12	Inspection chambers	)	
R12:13	Soakaways	)	
R12:14	Cesspits	)	
R12:15	Septic tanks	)	
R12:16	Preformed systems - to be measured in accordance with the rules of the appropriate section. Items R12:11-15.7 -13.1.0 are only measurable in non-preformed systems (i.e. building in ends of pipes, channels, benching, step irons, covers, intercepting traps and the like)		
R12:17	Connecting to Local Authority's sewer	B19.6	Connections to existing drains - refer GP8, work to be executed by a government or public authority
R12:18	Testing and commissioning	)	Not referred to in Section B19 but assume would be covered by P2
R12:19	Preparing drawings	)	
R12:20	Operating and maintenance manuals	)	

SMM7		POMI	
Clause	Heading	Clause	Heading/Comment
S	**PIPED SUPPLY SYSTEMS**	Q	**MECHANICAL ENGINEERING INSTALLATIONS**
T	**MECHANICAL HEATING/COOLING/ REFRIGERATION SYSTEMS**		
U	**VENTILATION/AIR CONDITIONING SYSTEMS**		
	All measured in accordance with the rules given under Section Y		Fittings to pipes of 60mm internal diameter or less are understood to be included. Fittings to pipes over 60mm internal diameter are enumerated but grouped together as "fittings"

SMM7		POMI	
Clause	Heading	Clause	Heading/Comment
V	<u>ELECTRICAL SUPPLY/ POWER/LIGHTING SYSTEMS</u>	R	<u>ELECTRICAL ENGINEERING INSTALLATIONS</u>
W	<u>COMMUNICATIONS/ SECURITY/CONTROL SYSTEMS</u>		
	All measured in accordance with the rules given under Section Y	R3	Cables and conduit in sub-circuits can be enumerated or measured out in linear metres
		R4	Cable and conduit in final sub-circuits (i.e. to termination points) are enumerated

\multicolumn{2}{c	}{SMM7}	\multicolumn{2}{c}{POMI}
Clause	Heading	Clause	Heading/Comment
X	**TRANSPORT SYSTEMS**	P	**CONVEYING SYSTEMS**
X:1	Lifts	)P1.1	Lifts, hoists,
X:2	Escalators	)	conveyors, escalators
X:3	Moving pavements	)	or the like
X:4	Hoists	)	
X:5	Cranes	)	
X:6	Travelling cradles	)	
X:7	Goods distribution	)	
X:8	Mechanical document conveying	)	
X:9	Pneumatic document converying	)	
X:10	Automatic document filing and retrieval. All to be enumerated and described in detail	)	
X:11	Marking positions	)P2	Sundries – supports,
X:12	Identification	)	identification,
X:13	Testing and commissioning	)	testing, commissioning, tools,
X:14	Temporary operation of installations to Employer's requirements	)	spares and documents
X:15	Preparing drawings	)P3	Work incidental to conveying systems
X:16	Operating and maintenance manuals	)	

SMM7		POMI	
Clause	Heading	Clause	Heading/Comment
Y	<u>MECHANICAL AND ELECTRICAL SERVICES MEASUREMENT</u>	Q	<u>MECHANICAL ENGINEERING INSTALLATIONS</u>
		R	<u>ELECTRICAL ENGINEERING INSTALLATIONS</u>
	The rules for Section Y are applied to work covered by Sections R, S, T, U, V & W		

SMM7		POMI	
Clause	Heading	Clause	Heading/Comment
			Items not specifically covered by SMM7
–	These items are likely to be dealt with by measurement under the rules of the Civil Engineering Standard Method of Measurement (CESMM2)	B23 B24 B25 B26	Railway work Tunnel excavation Tunnel lining Tunnel support and stablisation

APPENDIX A
List of SMM7 Work Groups/Sections relative to Chapter 3

		Page
General Rules		14
A	**PRELIMINARIES/GENERAL CONDITIONS**	
A10	Project particulars	20
A11	Drawings	20
A12	The site/Existing buildings	20
A13	Description of the work	20
A20	The Contract/Sub-contract	20
A30	Employer's requirements: Tendering/Sub-letting/Supply	26
A31	Employer's requirements: Provision, content and use of documents	26
A32	Employer's requirements: Management of the Works	26
A33	Employer's requirements: Quality standards/control	26
A34	Employer's requirements: Security/Safety/Protection	21, 22, 24-26
A35	Employer's requirements: Specific limitations on method/sequence/timing	21, 22
A36	Employer's requirements: Facilities/Temporary works/Services	21, 22
A37	Employer's requirements: Operation/Maintenance of the finished building	26
A40	Contractor's general cost items: Management and staff	24
A41	Contractor's general cost items: Site accommodation	25
A42	Contractor's general cost items: Services and facilities	23-27
A43	Contractor's general cost items: Mechanical plant	24
A44	Contractor's general cost items: Temporary works	24-26
A50	Work/Materials by the Employer	23
A51	Nominated sub-contractors	22, 23
A52	Nominated suppliers	23
A53	Work by statutory authorities	23
A54	Provisional work	26
A55	Dayworks	26
B	**COMPLETE BUILDINGS**	
B10	Proprietary buildings	*
C	**DEMOLITION/ALTERATION/RENOVATION**	
C10	Demolishing structures	28, 31, 32
C20	Alterations - spot items	30
C30	Shoring	28, 31, 32
C40	Repairing/Renovating concrete/brick/block/stone	34
C41	Chemical dpcs to existing walls	34
C50	Repairing/Renovating metal	35
C51	Repairing/Renovating timber	35
C52	Fungus/Beetle eradication	35

Page

D	**GROUNDWORK**	
D10	Ground investigation	*
D11	Soil stabilization	*
D12	Site dewatering	*
D20	Excavating and filling	36-41
D30	Cast in place concrete piling	43-49
D31	Preformed concrete piling	50-55
D32	Steel piling	56-60
D40	Diaphragm walling	61-63
D50	Underpinning	105-107
E	**IN SITU CONCRETE/LARGE PRECAST CONCRETE**	
E10	In situ concrete	64-70, 85
E11	Gun applied concrete	68
E20	Formwork for in situ concrete	69, 72-80, 85
E30	Reinforcement for in situ concrete	70-71, 85
E31	Post tensioned reinforcement for in situ concrete	86
E40	Designed joints in in situ concrete ...	68
E41	Worked finishes/Cutting to in situ concrete	68, 69, 70
E42	Accessories cast into in situ concrete	70
E50	Precast concrete large units	81-85
E60	Precast/Composite concrete decking	85
F	**MASONRY**	
F10	Brick/Block walling	87-104
F11	Glass block walling	87-104
F20	Natural stone rubble walling	108-112
F21	Natural stone/ashlar walling/dressings	113-121
F22	Cast stone walling/dressings	113-121
F30	Accessories/Sundry items for brick/block/stone walling	92, 96, 100-104, 176
F31	Precast concrete sills/lintels/copings/features	81-83
G	**STRUCTURAL/CARCASSING METAL/TIMBER**	
G10	Structural steel framing	167-170
G11	Structural aluminium framing	167-170
G12	Isolated structural metal members	167-170
G20	Carpentry/Timber framing/First fixing .	147-150, 160, 166
G30	Metal profiled sheet decking	129, 135-139
G31	Prefabricated timber unit decking	129, 135-139
G32	Edge supported/Reinforced woodwool slab decking	129, 135-139
H	**CLADDING/COVERING**	
H10	Patent glazing	260-261
H11	Curtain walling	11
H12	Plastics glazed vaulting/walling	260-261
H13	Structural glass assemblies	260-261
H14	Concrete rooflights/pavement lights ...	12
H20	Rigid sheet cladding	151-153
H21	Timber weatherboarding	151-153
H30	Fibre cement profiled sheet cladding/covering/siding	129-131, 133-135, 146
H31	Metal profiled/flat sheet cladding/covering/siding	129, 133-135
H32	Plastics profiled sheet cladding/covering/siding	129, 133-135
H33	Bitumen and fibre profiled sheet cladding/covering	129, 133-135

Page

H	**CLADDING/COVERING (CONTINUED)**	
H40	Glass reinforced cement cladding/features	81-84
H41	Glass reinforced plastics cladding/features	129, 133-135
H50	Precast concrete slab cladding/features	81-84
H51	Natural stone slab cladding/features	227-234
H52	Cast stone slab cladding/features	227-234
H60	Clay/Concrete roof tiling	129-133
H61	Fibre cement slating	129-133
H62	Natural slating	129-133
H63	Reconstructed stone slating/tiling	129
H64	Timber shingling	129-133
H70	Malleable metal sheet prebonded coverings/cladding	129, 141-145
H71	Lead sheet coverings/flashings	129, 141-145
H72	Aluminium sheet coverings/flashings	129, 141-145
H73	Copper sheet coverings/flashings	129, 141-145
H74	Zinc sheet coverings/flashings	129, 141-145
H75	Stainless steel sheet coverings/flashings	129, 141-145
H76	Fibre bitumen thermoplastic sheet coverings/flashings	129, 141-145
J	**WATERPROOFING**	
J10	Specialist waterproof rendering	214-219
J20	Mastic asphalt tanking/damp proof membranes	122-127
J21	Mastic asphalt roofing/insulation/finishes	122-127
J22	Proprietary roof decking with asphalt finish	122-127
J30	Liquid applied tanking/damp proof membranes	122-127
J31	Liquid applied waterproof roof coatings	122-127
J32	Sprayed vapour barriers	*
J33	In situ glass reinforced plastics	*
J40	Flexible sheet tanking/damp proof membranes	129-130, 139-141
J41	Built up felt roof coverings	129-130, 139-141
J42	Single layer plastics roof coverings	129-130, 135-139
J43	Proprietary roof decking with felt finish	129-130, 135-139
K	**LININGS/SHEATHING/DRY PARTITIONING**	
K10	Plasterboard dry lining	247-250
K11	Rigid sheet flooring/sheathing/linings casings	151-153, 157-159, 235
K12	Under purlin/Inside rail panel linings	151-153, 157-159
K13	Rigid sheet fine linings/panelling	157
K20	Timber board flooring/sheathing/linings/casings	151-153, 157-159
K21	Timber narrow strip flooring/linings	151-153, 157-159
K30	Demountable partitions	247-250
K31	Plasterboard fixed partitions/inner walls/linings	247-250
K32	Framed panel cubicle partitions	163
K33	Concrete/Terrazzo partitions	81-83
K40	Suspended ceilings	251

Page

K	**LININGS/SHEATHING/DRY PARTITIONING** (CONTINUED)	
K41	Raised access floors	12
L	**WINDOWS/DOORS/STAIRS**	
L10	Timber windows/rooflights/screens louvres	162
L11	Metal windows/rooflights/screens louvres	171-173
L12	Plastics windows/rooflights/screens/ louvres	162, 171-173
L20	Timber doors/shutters/hatches	161
L21	Metal doors/shutters/hatches	161, 173-175
L22	Plastics/Rubber doors/shutters/hatches	173
L30	Timber stairs/walkways/balustrades	162
L31	Metal stairs/walkways/balustrades	174
L40	General glazing	257-259
L41	Lead light glazing	259
L42	Infill panels/sheets	240
M	**SURFACE FINISHES**	
M10	Sand cement/Concrete/Granolithic screeds/flooring	214-219, 226
M11	Mastic asphalt flooring	122
M12	Trowelled bitumen/resin rubber-latex flooring	214-219
M20	Plastered/Rendered/Roughcast coatings	214-219
M21	Insulation with rendered finish	220
M22	Sprayed mineral fibre coatings	221
M23	Resin bound mineral coatings	214-219
M30	Metal mesh lathing/Anchored reinforcement for plastered coatings	177, 222
M31	Fibrous plaster	252-254
M40	Stone/Concrete/Quarry/Ceramic tiling/ Mosaic	236-239
M41	Terazzo tiling/In situ terrazzo	223, 236-239
M42	Wood block/Composition block/Parquet flooring	236-239
M50	Rubber/Plastics/Cork/Lino/Carpet tiling/sheeting	241, 244, 245
M51	Edge fixed carpeting	255-256
M52	Decorative papers/fabrics	262-267
M60	Painting/Clear finishing	262-267
N	**FURNITURE/EQUIPMENT**	
N10	General fixtures/furnishings/equipment	164, 175, 177
N11	Domestic kitchen fittings	164
N12	Catering equipment	175, 187
N13	Sanitary appliances/fittings	187
N14	Interior landscape	*
N15	Signs/Notices	166, 266
N20	)	164, 175
N21	) Special purpose fixtures/	164
N22	) furnishings/equipment	164
N23	)	164
P	**BUILDING FABRIC SUNDRIES**	
P10	Sundry insulation/proofing work/fire stops	165
P11	Foamed/Fibre/Bead cavity wall insulation	12
P20	Unframed isolated trims/skirtings/ sundry items	154-156, 174, 177
P21	Ironmongery	166

A/4

		Page
P	**BUILDING FABRIC SUNDRIES (CONTINUED)**	
P22	Sealant joints	13
P30	Trenches/Pipeways/Pits for buried engineering services	39, 189-190, 211-212
P31	Holes/Chases/Covers/Supports for services	191
Q	**PAVING/PLANTING/FENCING/SITE FURNITURE**	
Q10	Stone/Concrete/Brick kerbs/edgings/channels	83-84
Q20	Hardcore/Granular/Cement bound bases/sub-bases to roads/pavings	36
Q21	In situ concrete roads/pavings/bases	64
Q22	Coated macadam/Asphalt roads/pavings	224
Q23	Gravel/Hoggin roads/pavings	42
Q24	Interlocking brick/block roads/pavings	242, 243
Q25	Slab/Brick/Sett/Cobble pavings	225, 242, 243
Q26	Special surfacings/pavings for sport	246, 266
Q30	Seeding/Turfing	41
Q31	Planting	41
Q40	Fencing	174, 273-275
Q50	Site/Street furniture/equipment	84
R	**DISPOSAL SYSTEMS**	
R10	Rainwater pipework/gutters	181-189
R11	Foul drainage above ground	181-189
R12	Drainage below ground	268-272
R13	Land drainage	268-272
R14	Laboratory/Industrial waste drainage	182
R20	Sewage pumping	13
R21	Sewage treatment/sterilisation	13
R30	Centralised vacuum cleaning	182
R31	Refuse chutes	182
R32	Compactors/Macerators	182
R33	Incineration plant	182
S	**PIPED SUPPLY SYSTEMS**	)
T	**MECHANICAL HEATING/COOLING/REFRIGERATION SYSTEMS**	) All to be measured
U	**VENTILATION/AIR CONDITIONING SYSTEMS**	) in accordance with) the relevant clauses
V	**ELECTRICAL SUPPLY/POWER LIGHTING SYSTEMS**	) of Work Section Y) see pages 192-210
W	**COMMUNICATIONS/SECURITY/CONTROL SYSTEMS**	)
X	**TRANSPORT SYSTEMS**	13
Y	**MECHANICAL AND ELECTRICAL SERVICES MEASUREMENT**	
Y10	Pipelines	)
Y11	Pipeline ancillaries	)
Y20	Pumps	)
Y21	Water tanks/cisterns	)
Y22	Heat exchangers	)
Y23	Storage cylinders/calorifiers	)
Y24	Trace heating	)
Y25	Cleaning and chemical treatment	)
Y30	Air ductlines	) 192-200
Y31	Air ductline ancillaries	)
Y40	Air handling units	)
Y41	Fans	)
Y42	Air filtration	)
Y43	Heating/Cooling coils	)

Page

Y	MECHANICAL AND ELECTRICAL SERVICES MEASUREMENT (CONTINUED)	
Y44	Humidifiers ....)	
Y45	Silencers/Acoustic treatment ....)	
Y46	Grilles/Diffusers/Louvres ....)	
Y50	Thermal insulation ....)	
Y51	Testing and commissioning of mechanical services ....)	
Y52	Vibration isolation mountings ....)	
Y53	Control componenets - mechanical ......)	192-200
Y54	Identification - mechanical ....)	
Y59	Sundry common mechanical items ....)	
Y60	Conduit and cable trunking ....)	
Y61	HV/LV cables and wiring ....)	
Y62	Busbar trunking ....)	
Y63	Support components - cables ....)	
Y70	HV switchgear ....)	
Y71	LV switchgear and distribution boards .)	
Y72	Contactors and starters ....)	
Y73	Luminaires and lamps ....)	201-210
Y74	Accessories for electrical services ...)	
Y80	Earthing and bonding components ....)	
Y81	Testing and commissioning of electrical services ....)	
Y82	Identification - electrical ....)	
Y89	Sundry common electrical items ....)	
Y92	Motor drives - electric ....)	
	ADDITIONAL RULES - WORK TO EXISTING BUILDINGS	13

* No specific rules included - see General rules clause 11

APPENDIX B
List of CESMM2 Work Classes relative to Chapter 4

Work Classification   Page

Class A: General items 284-291
Class B: Ground investigation 292
Class C: Geotechnical and other specialist
         processes 292, 297-298
Class D: Demolition and site clearance 291, 292
Class E: Earthworks 292-294, 320
Class F: In situ concrete 299
Class G: Concrete ancillaries 300-304
Class H: Precast concrete 304-305
Class I: Pipework - pipes 321, 323
Class J: Pipework - fittings and valves 321, 323
Class K: Pipework - manholes and pipework
         ancillaries 323
Class L: Pipework - supports and protection,
         ancillaries to laying and
         excavation 323
Class M: Structural metalwork 309-310
Class N: Miscellaneous metalwork 309-311
Class O: Timber 311-312
Class P: Piles 294-297
Class Q: Piling ancillaries 294-297
Class R: Roads and pavings 316-318
Class S: Rail track 280
Class T: Tunnels 280
Class U: Brickwork, blockwork and masonry .. 306-308
Class V: Painting 314-315
Class W: Waterproofing 313
Class X: Miscellaneous work 280, 320
Class Y: Sewer renovation and ancillary
         works 280

APPENDIX C
List of POMI Sub-sections relative to Chapter 5

                                              Page

SECTION GP - GENERAL PRINCIPLES
GP1	Principles of measurement	326, 327
GP2	Bills of quantities	326
GP3	Measurement	326
GP4	Items to be fully inclusive	-
GP5	Description of items	326, 342
GP6	Work to be executed by a specialist nominated by the employer	329
GP7	Goods, materials or services to be provided by a merchant or tradesman nominated by the employer	329
GP8	Work to be executed by a government or public authority	329
GP9	Dayworks	329
GP10	Contingencies	-

SECTION A - GENERAL REQUIREMENTS
A1	Conditions of contract	329
A2	Specification	327
A3	Restrictions	329
A4	Contractor's administrative arrangements	329
A5	Constructional plant	329
A6	Employer's facilities	329
A7	Contractor's facilities	329
A8	Temporary works	334, 362
A9	Sundry items	329

SECTION B - SITE WORK
B1	Site exploration generally	333
B2	Trial holes	333
B3	Boreholes (including pumping test wells)	333
B4	Site preparation	333
B5	Demolitions and alterations	331, 333
B6	Shoring	331
B7	Underpinning	335
B8	Earthworks generally	333, 334
B9	Excavation	333, 335, 362
B10	Dredging	*
B11	Disposal	334
B12	Filling	335
B13	Piling generally	335
B14	Driven piling	335
B15	Bored piling	335
B16	Sheet piling	335
B17	Performance designed piling	335
B18	Testing piling	335
B19	Underground drainage	362, 363
B20	Paving and surfacing	359
B21	Fencing	355, 360
B22	Landscaping	359
B23	Railway work	*
B24	Tunnel excavation	*
B25	Tunnel linings	*
B26	Tunnel support and stabilisation	*

* - Not dealt with in SMM7

Page

SECTION C - CONCRETE WORK
C1	Generally	337-339
C2	Poured concrete	335, 336, 337
C3	Reinforcement	338, 339
C4	Shuttering	338
C5	Precast concrete	339, 342, 345, 346, 350
C6	Prestressed concrete	339, 342, 346, 350
C7	Sundries	337, 338

SECTION D - MASONRY
D1	Generally	340-342
D2	Walls and piers	340, 341
D3	Sills, etc	340, 341, 342
D4	Reinforcement	342
D5	Sundries	336, 342

SECTION E - METALWORK
E1	Generally	343
E2	Structural metalwork	337, 343
E3	Non-structural metalwork	351

SECTION F - WOODWORK
F1	Generally	343-344
F2	Structural timbers	343
F3	Boarding and flooring	344, 346
F4	Grounds and battens	344
F5	Framework	343
F6	Finishings and fittings	346, 357, 358
F7	Composite items	327, 343, 344, 345, 351
F8	Sundry items	344
F9	Metalwork	344
F10	Ironmongery	358

SECTION G - THERMAL AND MOISTURE PROTECTION
G1	Generally	342-348
G2	Coverings and linings	344, 345, 346, 347, 348
G3	Damp-proof courses	342
G4	Insulation	345, 346, 347, 348, 357

SECTION H - DOORS AND WINDOWS
H1	Doors	351
H2	Windows	351
H3	Screens	345, 351
H4	Ironmongery	351
H5	Glass	352
H6	Patent glazing	345

SECTION J - FINISHES
J1	Generally	348-356
J2	Backgrounds	348, 353, 354
J3	Finishings	349, 350, 353, 354
J4	Sundries	349
J5	Suspended ceilings	350
J6	Decorations	354
J7	Signwriting	356

SECTION K - ACCESSORIES
K1	Generally	349-356
K2	Partitions	349, 350

SECTION L - EQUIPMENT 356
SECTION M - FURNISHINGS 356

	Page
SECTION N - SPECIAL CONSTRUCTION .... *	* Not dealt with in SMM7
SECTION P - CONVEYING SYSTEMS	366
SECTION Q - MECHANICAL ENGINEERING INSTALLATIONS	356, 361, 362, 364, 367
SECTION R - ELECTRICAL ENGINEERING INSTALLATIONS	358, 365, 367